OK, I've Signed Up for Statistics. Now What?

A Student's Guide to Navigation and Success in Statistics

Barry Trunk and Leslie Olsen

University Press of America,® Inc.
Lanham • Boulder • New York • Toronto • Plymouth, UK

Copyright © 2016 by University Press of America,® Inc.
4501 Forbes Boulevard, Suite 200, Lanham, Maryland 20706
UPA Acquisitions Department (301) 459-3366

Unit A, Whitacre Mews, 26-34 Stannary Street,
London SE11 4AB, United Kingdom

All rights reserved
Printed in the United States of America
British Library Cataloguing in Publication Information Available

Library of Congress Control Number: 2016934540
ISBN: 978-0-7618-6759-3 (pbk : alk. paper)—ISBN: 978-0-7618-6760-9 (electronic)

∞™ The paper used in this publication meets the minimum requirements of American National Standard for Information Sciences Permanence of Paper for Printed Library Materials, ANSI/NISO Z39.48-1992.

Contents

Acknowledgments

Writing a book is never a solitary event. Whether one author works alone, two authors corroborate (as Leslie and I have done here), or several people team together, the wisdom and suggestions of friends and professional colleagues is always welcome. Our book is no different.

The authors would first like to recognize the institutions in which we teach. Capella University and Bellevue University have supported and encouraged our work, and we appreciate the warm and professional environments of these institutions of higher learning supply.

In particular, Ms. Susan Simiele, adjunct professor at Bellevue University, has taken the time to thoroughly read the book, finding errors that escaped our keen eyes, phrases that we thought were clear but needed refinement, and other small errors of typing (such as an extra space between two words!) that needed correction. In addition, her knowledge of statistics reinforced that the materials within the text were relevant and important for students to be familiar with in their meaning and interpretation.

Dr. Brian Zax, adjunct professor at Capella University, has also generously given of his time and expertise in statistics to review and validate not only the statistical information in the pages you are about to read, but the concepts and ideas that form the foundation of statistical analysis. In fact, we might suggest that we have strong inter-rater reliability between our two colleagues, which is always a good thing to have!

I am also most appreciative of my good friend, colleague and co-author of previous books published by the University Press of America, Dr. William Cooney, professor of philosophy at Florida Gulf Coast University, for finding numerous errors in phrasing that we were certain were not there! A very keen eye indeed for a philosopher!

Last, but certainly not least, Leslie and Barry wish to acknowledge and thank the (literally) thousands of students we have worked with over the years. It is because of the feedback, comments, criticism and suggestions from you, the statistics student, that the real impetus to write a book that was easy to read, informative, and helpful in alleviating anxiety was born. This book is for you, and we hope we have done you proud.

Introduction

This book is an effort on the part of two very different educators. One of us is a core faculty member in the School of Social and Behavioral Sciences at Capella University, and teaches statistics; the other one is an adjunct faculty member at Bellevue University, a writing specialist who has written hundreds of pages of material on academic writing in multiple disciplines. One of us works primarily with numbers, the other primarily with words. Both of us are committed to your success as a student in statistics.

It is probably more true than not that since the very first time the very first student in the very first statistics course sat down and started to listen to the very first lecture, he (or maybe it was a she?) had the very first anxiety response. Here are a few hypothetical thoughts that might have been running through his or her mind as the lecturer started to speak:

"What in the world is he talking about?"

"Am I the only one in this class who is confused?"

"Note to self: talk to mom and dad about reconsidering that offer to join the family business."

"Why didn't I pursue that major in art history? I'll bet there were no statistics required for that one."

OK now, be honest . . . have you yourself had similar thoughts as you sat down in your stat class? Or when you started flipping through your stat text book? Hmmmmm????

As one who has instructed thousands of statistics students from high school to graduate school, I have seen the initial anxiety and trepidation among many of my new learners (not all mind you, but many indeed). It is with this audience in mind that we offer this small book to you. It is not in any way meant to be a substitute for your required statistics text; however, we hope that the strategies, hints, tips, and other suggestions will make the journey through your upcoming course more enjoyable, more understandable, and more satisfying. Whether it be focusing on some of the more common statistical concepts, thinking about how to write that paper in a manner that truly shows professionalism in the presentation of statistical information, or just becoming more familiar with statistics and statistical writing (and hopefully, reducing any anxiety!), this book is for you.

We have tried to write the text so that students in both traditional face to face courses, as well as those taking on online statistics class can equally benefit. Additionally, while SPSS (the Statistical Package for Social Science, published by the IBM Corporation) is mentioned in discussions of statistical software throughout the text, it should be noted that many other fine programs are available for the actual computation of statistical results, and the production of statistical output.

While we have tried to have an overall tone of consistency throughout the various chapters of the text, there may be some sections of the book where one of our personalities seems to "peek out" at the reader. But we don't mind, and hope you will not either. The careful reader might be able to tell which "voice" is behind the writing, but on the whole it is the content, not the person writing it, that is important.

We want you to learn from and enjoy our work, but mostly we want you to relax and enjoy your next statistics course. Yes, we DID say enjoy your statistics course (that was not a misprint!), and we hope our words of encouragement and support will allow you to do your best in these challenging classes!

Please let us know if the book is helpful! We can be reached at btrunk526@aol.com and leslieolsen@gmail.com. We would sincerely enjoy hearing from you!

So, what do you say? Shall we give it a go? Alright then!

Chapter One

OK . . . Let's Get Started

You've heard that familiar phrase from a Radio Shack ad: "You've got questions, we've got answers," right? Well, if we modify and expand this basic idea, we might come up with this:

YOU: *I am a student, and I am signed up for Inferential Statistics. I have three classes to take, and I have several questions—not to mention some anxieties—that I would like to ask, but I am not sure who to ask, or if they are even good questions to ask.*

Sound like anyone you know? Hmmmmmm?

As for the second part of the Radio Shack phrase, we might make it look like this:

ME: As a professor of psychology who teaches courses in both undergraduate and graduate statistics, I am very aware of the questions and the anxieties that many students have as they enter into Inferential Statistics. Here are my best attempts at answering your questions and your concerns . . . .

So, does this sound like something that might interest you? Are you willing to plunk down a few of your hard earned dollars to read a book of questions and answers from students who have taken the course you are about to embark on, and answers supplied by an expert in statistics and an expert in writing? Do you think that advice from educators who have taught literally thousands of students, just like you, might be useful? We think so, and that is why we have undertaken to write this book in a fun, straight forward and humorous (at least we think our jokes are funny!) manner.

But don't be fooled by our cool, casual style; we have REAL answers to REAL concerns that you probably have about your statistical education. Without sounding glib, we have heard it all before (as have the other fine faculty who teach these courses), and we take your concerns very seriously.

We want you to not only succeed, but to truly understand and appreciate the role that statistical analysis plays in professional research. We want you to be able to "do" statistics, but just putting numbers into a statistical software package and generating results, or perhaps going "old school" with pencil, paper and calculator to find a result is not the same as truly understanding what you are doing, why you are doing it, and what it all means. Those, my friends, are the true goals of graduate and undergraduate statistics, and outcomes that move you forward in your journey towards your undergraduate or graduate degree.

As we hope you can see, our desire in this manuscript is to address your questions early on (perhaps BEFORE you take that first statistics class, or during your first term of statistics), to give honest, reasonable and time-proven advice and answers, and to decrease any unreasonable anxiety you may have about the class, or the contents within it. More than anything, we want this book to help you move forward with learning the concepts and procedures, applying the materials, understanding statistical analysis, and then explaining your understanding to other academics through clear, concise writing. Our goal is to reduce your statistical anxiety, to help you stop worrying about the class and unknown aspects of undergraduate and graduate level statistics. This book attempts to make the unknown known, the mysterious understandable, and the heretofore enigmatic concepts of statistical analysis accessible. Of course, there is no substitute for quality study time, and your work in statistics will likely use up a pretty good chunk of your overall time spent hitting the books. But believe us when we tell you it is well worth the investment, as the study of statistics transcends all fields of social, physical and biological science, and helps you to develop and hone your critical thinking skills.

Please keep in mind that this book and the material it contains is NOT meant to replace your required statistics text book, or any information given to you by your instructor. Rather, this book should be looked at as a useful supplement to your text, as an easy to read and easy to follow resource to topics and concepts introduced in your course, and as a reminder to believe in yourself and your abilities.

Now, since I (Barry) am a member of SOSBS at Capella University (School of Social and Behavioral Sciences) I can only speak for courses in that particular School. I have not taught statistics for other disciplines, such as business, biology, or astronomy. Does this mean that students in those schools, or with majors other than psychology would not benefit from this book?

Not at all! Your money is most welcome (uh, what I mean is that students in Schools outside of SOSBS, as well as other universities, colleges or high schools are just as likely to have the same questions and concerns that SOSBS students have, and the guidelines and advice given here are just as appropriate to those taking statistics in the other disciplines). Additionally, I (Leslie) have taught writing about statistics in multiple disciplines, including business, technology, education, and nursing, among others. The point to take away is that, in the end, a mean is a mean, a standard deviation is a standard deviation, and statistics is statistics. The *content* may be different, but the *concepts*, *procedures* and *interpretations* of the data are the same.

OK, we seem to understand each other so far (you are still reading this, which we take as a positive sign). To better gauge where you might be on the statistical roadmap, take a look at the following three scenarios, and see which one is closest to describing you:

SCENARIO 1

You are filled with wonder and awe as you begin your study of inferential statistics. You dream of incorporating research design with parametric (and non-parametric!) statistical techniques, distinguishing between sample and sampling distributions, exploring the domains of probability, linear and multiple regression, and eagerly await your inspiration for your senior paper, Master's thesis or doctoral dissertation. You can hardly sleep, anticipating next week's reading on t tests. The very mention of the Analysis of Variance makes you shudder with excitement. When you do finally fall asleep at night, you experience the sweet dreams of advanced techniques such as Repeated Measures ANOVA, Logistic Regression, Factor Analysis, Discriminant Function Analysis . . . ah, sweet bliss . . . .

Is this you? Yes? No? How about this one:

SCENARIO 2

You are dreading the course. You have put the class off as long as you possibly can, and now you have no choice; you MUST take it. You would rather get a root canal than learn about correlation. You hate numbers, and numbers hate you. Your last calculator was free with the purchase of three rolls of Scotch tape. Statisticians rank just slightly below used car salesmen, and you are sure that SPSS really stands for something besides Statistical Package for Social Science. You hire someone else to count your sheep for you at night, and if you could buy your way out of this, you would consider it! All those tables, charts, numbers, formulas and graphs! Qualitative analysis is looking better all the time (with apologies to my good friends, Dr. Bill Percy and Dr. Kim Kostere). You feel there is just no way you are going to do well, because you are not a "numbers person."

So, is this you? Yes? No? Well, how about this one:

SCENARIO 3

You are somewhat apprehensive about the course. You are OK with numbers and statistics, but complex formulas and difficult conceptualizations make you nervous. You are not sure about using a statistical software package for doing anything except the most elementary and basic of calculations, and the thought of doing problems by hand makes you nervous. Perhaps your statistics is rusty, as you had a class several years ago and got a B in it. Or perhaps this is the first time you have had to take an actual statistics class, and heard "rumors" that it is really hard and confusing. You want to do well, hold your own, and learn as much as you can about what is typically a challenging subject. You have a good attitude, realize that the class involves a fairly large time commitment, but are ready to give it your best.

So, where are you on the hypothetical spectrum of statistical motivation? Which of the three scenarios seems best to fit your feelings and personality? Hopefully, most of you are in the middle (Scenario 3); a bit skeptical, a bit apprehensive, but ready to give it a go. We want to emphasize that we fully understand your feelings. But we also know you would not be here if you were not ready to be here. The fact that you have gone through your program, whether it be high school, college, or graduate school, and come this far speaks volumes about your character, your determination, and your persistence. Remember, this is just another course on the journey to your degree, and not only will you learn a great deal, you may actually find it to be enjoyable as well. I (Barry) have taught the course for many years now, and most students not only do very well, but are genuinely pleased and surprised at how much they understand and can apply to their own thesis or dissertation studies. And I (Leslie) have help hundreds of students with their writing skills and presentations of everything from a short paper to a dissertation.

So, where do we go from here?

Asking questions and receiving answers is very familiar to all of us. Since during our years of teaching we have encountered most of the questions you are likely to have, and from experience have answers that seem to have been successful for students just like you, it seemed natural to have the book laid out in an easy to read, easy to follow and easy to understand Q&A format. There may be other questions on your mind, but we have tried to anticipate most of them and keep them in a logical order (and remember, we love emails, and will answer you when we can!). Keep in mind that the guidance we give here is ours and ours alone; we do not claim to speak for any of the other fine faculty who teach statistics or write about statistics (but in discussion with them, we do feel that many would echo the ideas and suggestions we will be offering to you).

Ready? Sound good? Then pour your favorite drink, grab a comfortable chair, take a deep breath and let's go . . .

Chapter Two

Questions and Answers

Before the Class Starts

As promised, we begin with some Q&A to address your questions, starting with the anxiety-filled confessions of others who have been where you find yourself now, facing a course in statistics and statistical analysis. After calming your fears (we hope!), we move on to some of the more practical questions you will ask once you accept that you, too, can learn (and appreciate) statistics.

I Am Soooo Nervous!

YOU: *I'm really nervous. I haven't had statistics in quite a while, and when I did take it I really didn't understand it very well. I think everyone else will be smarter than me, and this will be embarrassing.*

ME: I hear this a lot. In fact, I would say that about 25% to 50% of people signing up for my classes say something like this in the class café, or in the introductory discussion posting. So the first thing to realize is that if you feel this way, you are not alone!

But, what can I tell you to make you feel better? To begin with, let me share some remarks from people who have just finished the class you are about to take. These are their actual words and comments:

Initially, my anxiety level was very high. I was anxious about grasping SPSS and interpreting the data results. My only goal for the class was to PASS! However, I am excited to say, that not only did I accomplish my goal of passing the class, I learned a lot of valuable information during the process. As a result of the help received from Dr. Trunk (you're the best!), and my classmates, I am able to analyze SPSS outputs with ease. This ease will definitely assist me with the results section of my dissertation. The most significant information learned in this class was learning the various data tests (ANOVA, chi-square, etc). This information opened my eyes as to different ways to analyze my dissertation data.

The quality of class participation and class discussions was top notch. I was able to learn from each student's post, as their postings help me to better understand some of the statistical terms and concepts. For future students, I would recommend taking Dr. Trunk's class, staying on top of the assigned readings, and start early on the application assignments. While this class is a lot of work, it is manageable if you manage your time wisely.

When this class first started, I felt like a deer looking in the head lights of an oncoming vehicle. As the class continued, I became a lot less frightened and a lot more interested in the various measures and designs that could be applied to my research. The biggest benefit from this course was my improved ability to understand and properly write up data. I truly went over and beyond my personal expectations for the class. I did not think I would do as well in understanding or grade but I was happily mistaken. The class discussions were awesome and the students in the class were very supportive. I would recommend students to find a teacher that is considerate and caring. I'm really glad that I did not have to look too far.

Yeah, I know . . . sounds too good to be true, but these are really pretty typical of the comments I receive after the class has ended. Did you notice, by the way, that the students note how much they learn from and with each other throughout the course? You have an automatic support system built into your course—tap into it!

Now, I am fully confident that all of the exceptional faculty who teach statistics have students who walk away from their courses with similar statements. So the lesson learned here is that hard work pays off, that you are probably in a much better situation to learn the material than you were earlier in your life, and that you too will succeed. Keeping an open mind, having exceptional time management, and a healthy, positive attitude will go a LONG way in these courses!

YOU: *Those are some pretty impressive comments. What about the bad comments?*

ME: What? What about the "bad" comments? There weren't any! No, really. I couldn't find any at all . . . go figure . . .

How Do I Prepare for Class?

YOU: *OK, I guess I feel a little better about the class, but I am still really unsure about myself and my role in it as a student. What do you suggest I do before the class actually starts?*

ME: This is an excellent question! The time to begin is before you begin, if you know what I mean. Here are some concrete suggestions:

1) Become intimate with the course syllabus. When you have access to the class, print (yes, print) out the syllabus. Now, read it thoroughly, then read it again. You would be amazed at how many questions we as instructors

get from students that are answered right there in the syllabus! We fully realize you are busy, professional adults, and your time is limited. But you must take the initiative to read what we give you. Yes, there is a lot of information here, and things do slip by occasionally, but you will be much better prepared to do well in the class if you understand what has been assigned, where it is located, when it is due, and so on.

2) Buy and install the latest version of statistical software. Use the package you're your particular school recommends (SAS, EXCEL, and SPSS, the Statistical Package for Social Science. are probably the most popular). Make certain that you order the correct version (see the syllabus!) for your course. Sometimes renting the software makes sense, while at other times buying it outright might be a better option. Keep in mind that for some statistical software, the student version is a barebones, stripped down program that does not contain all of the necessary components for graduate level statistics. In addition, the student version is usually limited to a small number of cases, and would likely be insufficient for your own statistical analysis of your thesis or dissertation.

3) License the software immediately after installation. If you do not do this, it WILL stop working after two weeks (that's a promise!), and you will be tempted to chuck the entire class and watch reruns of Gilligan's Island on TV. This is a major hassle, especially if you have waited until the second week, and your first assignment is due in four days. In order to license the software, you will need access to the Internet and the license code that comes with the product. Do this right after you have installed the software; do not wait! Here is a typical example of a student who didn't follow my golden instructions.

HELP! I am desperate! My SPSS just stopped working and I don't know why. I installed it and it was working just fine, but now I am getting some kind of weird licensing message. What should I do?

The question should be "what should I have done?" Please do not be one who writes this message to your instructor; buy the correct product, install it, and license it. Now go on and live a happy life (statistics wise, anyway).

YOU: *OK . . . so what can I do if I, um, forgot to license my software? Is all lost?*

ME: SPSS HAS A TECHNICAL SUPPORT NUMBER THAT STUDENTS CAN CALL FOR ASSISTANCE. There is also a SPSS website that answers questions about licensing authorization codes. Other software packages have similar options, but again, your life will be much easier if you simply license the software when prompted during installation.

4) Be sure you have the correct text book for the class. If your book is the wrong edition, I would suggest getting the right edition. I know, more money, but now you have the correct page numbers, the correct chapters, and the latest information. In short, it is money well spent. Now, that is not to say

you can't pick up a good used copy of the correct edition, or that you can't purchase a copy from a fellow student who may not need the book (having mastered the material themselves!). One tip . . . it is usually best to order via the university bookstore. That way you are sure of getting the correct edition and for the correct country (yes, there are international editions out there)!

JUST HOW MUCH MATH DO I NEED?

YOU: *I am not good at math. Do I need an expensive calculator? Do I even need a calculator?*

ME: Let me answer the second question first: Yes, you should buy a calculator (or borrow one from your kid). Does it need to be expensive? Well, that depends on what you consider to be expensive. My recommendation is to get a TI-30, which, last time I looked, was about $15.00 at a local discount store. A basic calculator that has the usual functions is perfectly adequate. Even an inexpensive calculator, like the TI-30, has many additional keys, such as logarithms, trigometric functions, and a memory. As always however, refer to your syllabus and purchase the make and model your instructor requires.

In some basic calculators, there are also some statistical keys, such as the sum of X and the sum of X squared, that are really useful for hand calculation of the variance of a distribution. In fact, if you have kids in high school, they may have a perfectly good calculator that you can use (hey . . . didn't I just say that a minute ago? SENIOR MOMENT)!

For you big spenders out there, a multitude of statistical calculators are available. Both Texas Instruments and Hewlett Packard make very good models. For many of you in an online or traditional class, you will most likely be using your computer and your SPSS (or other statistical) software. But for those who enjoy doing things by hand, seeing what all those little keys do, or just want to impress your friends, these calculators should more than do the job.

YOU: *OK, OK, I get the message. But I'm not very good at math. Can't I just use my software to do the problems?*

ME: Time for a bit of history. Back in the old days, students did most ALL of their problems by hand. And do you know who those "old" students are? They are the men and women who are teaching you statistics now! So, don't be at all surprised if they want you to do at least some of the simpler problems by hand, with your trusty TI-30. And most of these problems involve nothing more than ordinary arithmetic, and perhaps a bit of basic algebra.

And here is another little secret: it is kind of satisfying, in a weird geeky sort of way, to do a problem by hand, then put the data into SPSS and get the same result! Man, Maslow would probably classify this as a self-actualizing experience! You have arrived! Whoo hoo! Try it and see.

Now if this wasn't enough, there is another good reason to do a problem by hand: you will understand it so much better when you "see" where all the numbers are coming from. Many books go through simple calculations in a step-by-step manner. Try to follow along! Once again, most times the math is nothing more than addition, division, and squaring!

YOU: *I heard that statistics involved a lot of math. Is this true?*

ME: OK, I can see this is really a big concern. So let me tell you something that is the honest truth; the conceptual aspects of statistics are more difficult than the mathematical ones covered in most basic statistics courses.

YOU: *Conceptual? You mean abstractions and things like that?*

ME: Well, in a way, yes.

YOU: *How about an example?*

ME: OK. Here is one for you. Most of the time, we look at our world and examine things in a positive sense. Say you are in the market to buy a new oboe (oboe, not elbow. If you are in the market for a new elbow, please see my other publications, (Buying a New Elbow for Dummies). You find one you like, and buy it with the assumption that it is a good one. In other words, you look at it in the positive way (it is a good oboe), and keep that perspective unless you have reason to discard it (if the keys of the oboe fall off, you realize it is a bad one).

BUT, in statistics, we a form a special hypothesis (called the null hypothesis) that is framed in the negative. In the oboe example, we might assume it is a bad oboe, and then use it to see if it really is a bad one or not. If it seems to work OK, we "reject" our idea it is a bad oboe, and conclude it is a good one. See the difference?

This is just one of the many ways that statistics is conceptually difficult. And did you notice that there was not a single number in that example?

YOU: *OK, that was a good example. But c'mon, this is statistics, and statistics involves math!*

ME: Touché. But the actual mathematical calculations in some (but not all!) courses is minimal, and anyone who can do very basic operations (addition, division, multiplication, squaring) can try the more simple examples that are always given in most chapters. You can also make up a very simple set of data, and then solve the problem by hand.

I will say that in more advanced statistics books, derivations and calculations become more involved, but even at this level most students in the social sciences and other, non-mathematical disciplines, will not need to understand the mathematics to confidently perform the procedure, and offer reasonable interpretations of the results. So bottom line, the math is there for those who want to explore, but is not something you need get hung up about. Most will not need nor use the math, and still be quite able to run complex procedures, provide reasonable interpretations and inferences of the results, and read and understand the statistical information in the professional literature. And for many of us, that is more than enough!

WHAT ABOUT HOMEWORK?

YOU: *I know I am going to have homework to turn in. What are some of the reasons you as an instructor take off points?*

ME: Students often wonder why points were taken off from their homework. Here are some possible reasons:

1) Wrong answers were given. Check and double check your work; was the data inputted properly? Was the procedure followed exactly?

2) Brief and superficial responses (See below).

3) Wrong questions answered. Be careful here! Please be sure to send in ONLY the problems that have been assigned. Answering problems not assigned, or changing the order of the questions when you answer them will probably not go over well with your instructor.

4) Incomplete assignments (not all questions answered). This is a BIG one! For most assignments, you MUST include your SPSS output, copied and pasted onto your WORD document. It is not usually necessary or required for you to send SPSS files (but check with your instructor, as he or she might want them as well).

5) Assignment turned in late (See the syllabus, as different instructors might have different due times or dates). And if a real situation does arise, notify your instructor as soon as you are aware of it so he or she can make accommodations.

6) SPSS tables were required, but none were given. (This is a BIG one! See # 4 above).

7) SPSS tables and output were included, but not discussed. For myself, this is the main reason people will lose points. I need to know that you understand what you are producing. It is one to MAKE a table, but another to INTERPRET what it means. Remember, tables, charts and other outputs are not meant to stand alone. You cannot just make them, then say to the reader "OK, here they are . . . have fun."

As far as what I personally expect (for both discussion posts, feedback to students, and homework), a quality answer should be concise but complete. All your instructor has in front of them is what you send. Do not make assumptions such as "he knows I know this." Explain what you know, using appropriate and professional language (see Chapter 3 for more tips and discussions of quality writing).

YOU: *How about an example for us?*

ME: Sure . . . glad you asked. Here is an example from two students over a year ago, both answering this question:

"If the variance of a distribution is 64, what is the standard deviation?"

Student A answers: "8."

Student B answers: "Since the standard deviation is the square root of the variance, the correct answer is 8. Also, since standard deviations cannot be less than zero, the answer cannot be -8, even though this also is a root of 64."Now, if YOU were the instructor, which answer would impress you more? Would you give the same number of points to both students? Would you assume that Student A and Student B have the same level of understanding of the relationship between the variance and the standard deviation?

Your answers do not have to be overly long, but you need to show me (your humble instructor) what you know, and say a sentence or two, about the result. Don't just give a number with no explanation and expect to get many points for that.

An acceptable answer to the above simple question would be "The standard deviation is defined as the positive square root of the variance. The square root of 64 is 8, so the standard deviation is 8."

At the point of being redundant, for questions that ask for SPSS, you MUST copy and paste the proper output into your homework in the correct location. Also, VERBALLY discuss the answer! It is NOT enough just to insert a table and think you have answered the question. This applies to any final project, thesis or dissertation as well. You must discuss your results! And, those write ups must be in proper APA format. They need not be long and wordy, but must follow the well established format of the profession (See my colleagues' tips on quality writing later in Chapter 3)!

A LITTLE ABOUT VARIABLES

YOU: *I understand that there are two basic kinds of variables. Can you tell me a little more about them?*

ME: Sure! Glad to help out! Generally speaking, there are TWO kinds of variables, qualitative and quantitative. They must be analyzed and interpreted in TWO different ways.

There is always some confusion on the distinction between quantitative and qualitative variables. This is a central designation between variables, and determines the kinds of analyses that are permissible.

Quantitative variables have scores, numbers or other meaningful measures that are numerical. Many common measures, such as height, weight, GPA, annual income, number of babies born in a certain hospital, calories consumed in a week, and hundreds of others are variables where the numbers represent real, meaningful units of measure. It is perfectly acceptable to calculate statistics on quantitative distributions, including the familiar mean, variance, and standard deviation.

Note that numbers that refer to ranks, or attitudes, are not considered to be "real" in the sense of having the mathematical properties (scales of measurement) needed to do most parametric statistical analysis. This includes Likert type scales (On a scale of 1 to 7, how much do you like diet Coke)? Let's talk about this now.

Qualitative variables, in general, refer to categories of some sort, usually in the sense of comparisons. We might want to compare males to females, or Chevy's to Fords, or Online Universities to Traditional Universities, or different regions of the country (North, South, Midwest, East Coast, West Coast), and so on. Qualitative Variables are "dummy coded" with 1's and 2's, or some such system, purely to IDENTIFY the kind of variable (1 = males, 2 = females). This lets your software (like SPSS) count the number of each gender, and to find percentages. But, it is NOT correct to find the "mean" of gender (even though the software may well do this, if you ask)!

YOU: *Can you elaborate a bit?*

ME: Let me elaborate a bit. Say we code males as 1, and females as 0. This allows SPSS to count how many males and females are in the sample of data.

Pretend we have the following set of scores for gender:

{0,0,1,1,0,0,0,1,0,0,1,1,1,0,0,1}

Your software will "know" that there are 7 males and 9 females. Great. You can report this in a table, or find percentages.

But, to say that the "mean" of gender is 0.4375 is completely meaningless. So, the point is clear: you should not calculate statistics such as these for qualitative variables. Instead, you will find frequencies and percentages (what percent of the total sample is female? How many males are in the sample?). These kinds of questions are meaningful, and you will generate tables (such as bar or pie charts) that answer them.

YOU: *This makes sense, but don't we use qualitative AND quantitative variables together in most research?*

ME: This is another excellent question, but the answer is not quite black and white. While there are qualitative and quantitative variables, as we have just seen, there are also studies that are qualitative, and studies that are quantitative. We study these in some detail in classes devoted to research methods. For now, it is sufficient to say that qualitative studies are primarily non-numerical, involving things such as interviews, case studies, and lived experiences.

On the other hand, a quantitative study can, and often does, have one or more qualitative variables. For instance, we might be interested in how people of different religions (a qualitative variable) differ in terms of the amounts of money donated to charity (a quantitative variable). This is a quantitative study, but has a qualitative (categorical) variable in it.

Now that we have had some time to get acquainted, let's move on to Chapter 3 and continue the fun (fun = learning + understanding)!

Chapter Three

Some Things You Really Need to Know about Statistics!

OK. In Chapter 1 of the book, we looked at some of the things that typically cause anxiety in some students taking statistics; in an informal, non-scientific poll, I have found that the majority of students 1) are nervous about their upcoming course, 2) have not taken statistics for several years, and 3) are still worried about the math involved. But, having read the two chapters, you are now happy, comfortable and ready to take on the world (or at least take on that first stat course, right? RIGHT?).

In this chapter, we are going to look at some of the VERY BASIC things that you should be familiar with entering into the class. DON'T FREAK OUT! These are things that you most likely have seen before, but may not have been explained as well as they might have been. Let's take a look at the big picture first, then get into some of the details, OK?

Statistics is a branch of mathematics, but with the goal of the analysis and interpretation of data. Certainly math, equations, graphs, tables and figures are all important in statistical analysis, but with the advent of modern statistical programs (such as SPSS, the Statistical Package for Social Science), the actual computation of all but the most elementary of mathematical expressions is largely forgone. Sure, some diehards might still enjoy inverting matrices by hand, but for the rest of us we are happy to let the program do the math, and we can settle with the challenging aspect of what the heck it all means!

Having said that however, there is still merit in performing some of the easier analyses (and some not quite so easy) by hand. Doing the intermediate steps and calculations can, for some, really be enlightening, as one discovers the logical steps involved in the final outcome of an analytic procedure. So don't be afraid to tackle some of this with your trusty calculator; you might be surprised at how good it feels!

At this point, we want to familiarize you with what you can expect in the typical statistics courses you will be taking. Turns out that for many of these courses, the procedures fall into one of the five major categories listed below . . . hang on guys, here we go!

THE FIVE MAJOR CLASSIFICATIONS OF STATISTICAL WORK

I. Descriptive Statistics: This category involves the simple summarization of numerical data. These are displayed in tabular or graphic form. Common statistics such as measures of central tendency (mean, median, mode), measures of variability (standard deviation, variance, inter-quartile range), and measures of shape (skewness and kurtosis) are usually familiar from previous undergraduate classes. In addition, simple graphs such as histograms, scatterplots, and box and whisker plots "show" how the data look. Many of the graphs can superimpose a normal distribution over the output, so that one can quickly see if the data are roughly normal (bell shaped) and in the case of scatterplots, linear (approximately straight lined with one another).

II. Comparison of Groups: This category includes both simple univariate group comparisons, and in more advanced courses, multivariate group comparisons. All of these procedures basically look at the means of different groups, and asks if the differences between the means are large enough to be statistically significant (observed differences are probably not due to chance). Included here are

t tests—comparing one independent variable with exactly two groups on one quantitative dimension.

One way analysis of variance (ANOVA)—comparing one independent variable with more than two groups on one quantitative dimension.

Two way (factorial) ANOVA—comparing two independent variables on one quantitative dimension.

Repeated measures ANOVA—comparing scores of the same individuals across different time periods.

Mixed ANOVA—which has both a repeated measures (within subjects) variable, and one or more independent (between groups) variables.

Analysis of covariance (ANCOVA)—which makes an adjustment to each quantitative score for every participant, based on the nature of the correlation between a covariate (a quantitative variable correlated with the outcome dependent variable) and the dependent variable. After this adjustment (statistical control), ANOVA is run on the adjusted scores.

Note that all of the above have only ONE dependent variable. The following procedures have MORE than one dependent variable, and are considered multivariate methods:

Hotelling's t squared—a procedure similar to a t test where two groups are compared. The difference is that there are two (or more) dependent variables, which have been "combined" into a new, single multivariate outcome measure that incorporates attributes of both of the original variables simultaneously.

Multiple analysis of variance (MANOVA)—identical to Hotelling's t squared but having more than two groups in the independent variable.

Multiple analysis of covariance (MANCOVA), identical to MANOVA but with an adjustment of the dependent scores based on a correlated linear relationship to one or more quantitative covariates.

Factorial MANOVA—Similar to MANOVA but at least two independent variables are included.

Factorial MANCOVA—Similar to factorial MANOVA, but with the addition of one or more covariates.

III. Prediction of Scores and degrees of relationship between variables: Techniques and procedures used here are focused on discovering (most commonly) linear relationships between two variables (bivariate analysis) or more than two variables (multi-variate analysis). Correlation refers to both the degree (magnitude) of the relationship, and the direction (positive, negative, or zero). Regression refers to the prediction of a quantitative dimension (score) based on values from other predictor variables. Here are some examples of these procedures:

Bivariate correlation (r)—This method involves the calculation of the correlation coefficient between two continuous, quantitative variables. The correlation is a derived value that will always be between -1 and +1. The closer the value is to either of its limits, the stronger the relationship is. Conversely, the closer to zero, the weaker the relationship. Correlations are not to be inferred as causal; Variable A does not cause Variable B to occur. Rather, Variables A and B are related to one another with a certain level of strength. Even if two variables are perfectly correlated, we cannot legitimately infer any causal connection.

Bivariate regression—This procedure predicts the value of one variable (usually called the criterion variable, sometimes incorrectly called the dependent variable), based on the correlation it has with another variable (usually called the predictor variable, sometimes incorrectly called the independent variable). The outcome is the generation of a regression line that has a specific slope and y intercept (remember your old high school geometry?). This line is the best fitting line to the original data, and can be superimposed on a scatterplot. For a bit of an abstract understanding, there are an infinite number of straight lines that can be placed in a scatter plot, but there is only ONE line that at the same time minimizes the squared deviations from that line. That special line is called the regression line.

Multiple Regression and Correlation (R)—Similar to the above, but with more than one predictor variable. Multiple regression involves two or more predictor variables, combined, to generate a predicted value on the criterion variable. Most often we are interested in 1) how well the model (the predictors) account for the observed data (R squared), as well as how each individual predictor, independent of the others, correctly predicts the outcome measures. Multiple correlation symbolized by R) is the bivariate correlation of the original scores with the predicted scores.

Hierarchical (Sequential) Multiple Regression—A regression model where the researcher determines the order of entry of the predictor variables into the analysis. Contrast this method of analysis to the above, where all variables are given equal weight and entered at the same time.

Stepwise Multiple Regression—A regression model where the computer uses an algorithm to determine the order of entry into the analysis. Usually the variable with the highest bivariate correlation to the dependent measure is entered first, followed by the next highest one, and so on.

Canonical Correlation—In this multivariate method, there are several criterion variables and several predictor variables. Contrast to bivariate or multiple correlation, where there is only ONE criterion variable.

Chi Square Analysis—This is a fairly simple procedure that many undergraduate courses teach. The idea is that there may be a relationship between two categorical variables. For example, is there a relationship between political party (Democrat, Republican, Independent) and pet ownership (Dog, Cat, Bird, Other, None)?

Multiway Frequency Analysis—This is similar to chi square, except that a third and possibly fourth categorical variable is added. In the above example, we might add Gender and Religion as two additional variables.

IV. Prediction of Group Membership: In the above discussion, Regression is used to predict a quantitative dimension (score) based on two or more predictor variables. However, sometimes the goal of the analysis is to predict, or classify, an individual into a particular group, based on scores for a set of predictor variables. For instance, one might want to know if a person up for parole might re-offend (one group) or not re-offend (a second group). Or, we might want to know if a person undergoing heart surgery is likely to survive (one group) or not survive (a second group). Here are two methods that can be used for these research questions:

Discriminant Analysis—This method, also sometimes called discriminant function analysis, is used to predict group membership (the dependent variable) when a set of scores on predictor variables are given. The predictor variables are quantitative scales (for example, IQ scores), not categories (gender).

Logistic Regression—Logistic regression is used to predict group membership when some of the predictor variables are quantitative, and some are discrete. Logistic regression calculates the odds (probability) of being in one group or another when variables are a mix of continuous and discrete scales. Logistic regression is a non-parametric procedure, involving a transformation of the odds ratio into the logit (natural log of the odds ratio).

V. Analysis of Structure: These methods look at the underlying dimensions within a set of variables. Often the goal is to see if variables can be described by a higher order of magnitude, sometimes called a factor, or a latent variable. It is also usually of interest if the variables can be combined, or grouped together, into a much smaller set of factors, without significant loss of information. The search for structure can be empirical or theoretical.

Components Analysis—The goal here is to understand the relationships between variables in a correlation matrix. The large number of variables in the matrix are combined in a way that similar items are grouped together into components. Components analysis is descriptive, not theoretical.

Factor Analysis—This method has many similarities to components analysis, and is often discussed in the same chapter. There are some mathematical differences between the two, but both seek to reduce the number of observed variables into a much smaller subset of factors. Unlike components, factors are more involved in the search for theoretical meaning of the data, rather than a description of the relationships themselves.

Structural Equation Modeling—This method involves a combination of factor analysis and multiple regression. Some of the variables may be directly observed, while others may be latent. Variables may be used in combination to predict a quantitative dependent variable.

You: *Man . . . that was a really good review! If only you had included some important terms and concepts for me!*

Me: Well, since you asked . . .

SOME IMPORTANT TERMS AND CONCEPTS

- Measures of Central Tendency are statistics that try to capture the average, or typical aspect of a distribution. We use these all the time! What is the average price of gas? If you went bowling and bowled three games, what was your average score? If 400 participants take a test of creativity, what was the most typical score? As it turns out, there are several ways to answer the question of an "average" score, and as you might expect, different calculations can lead to different interpretations about the data!

The MEAN is just the arithmetic average of a set of scores. If all of us stepped on a scale, and we recorded all of our weights, we would find the mean weight by adding up all the scores and dividing by the number of scores. The MEDIAN is the middle score (when arranged from lowest to highest), and the MODE is the score, or scores, that occur most often in the distribution (note there will always be a mean and median, but not always a mode).

-Measures of Variability are statistics that tell us something about the dispersion of scores. Are most of the scores bunched closely together, or are they spread apart. Sets of scores (distributions) that have large variability have scores that are widely dispersed around the mean (average) score, while those with lower These are independent of the average score, and also can be calculated in different ways.

The RANGE of a distribution is a simple measure of its dispersion, or variability. Simply take the highest score and subtract from that the lowest score, and you have your range. Can you see that the minimum value of the range is 0 (can you say why)?

More useful is the STANDARD DEVIATION, which unlike the range uses ALL of the scores in its calculation, not just the lowest and highest. The Standard Deviation a kind of average of every persons score from the mean. Notice if all the scores are bunched together, they are all close to the mean, and the standard deviation will be small. If the scores are "all over the place," meaning some are very large and others are very small, then the standard deviation will be large. Like the range, the minimum value for the standard deviation is zero (why?).

- Measures of Association usually involve some form of correlation. Correlations look at if two (or more) variables are related in some way. One common example might be eye color and hair color. While there are certainly exceptions, many people with blond hair have blue eyes, and many people with dark hair have brown eyes. So it seems that eye color and hair color are in some ways related to one another! No big surprise here, there are many different ways to calculate a correlation!

Bivariate correlation involves two variables, and (not surprisingly) multiple correlation involves more than two variables. The former is calculated using the Pearson correlation statistic (r), while multiple correlation is calculated using the multiple correlation coefficient, R. Correlations of 0 indicate no linear relationship, while correlations of 1 indicate perfect linear relationship.

- Linear Regression refers to some form of prediction. Since correlations look at relationships, they are closely related to regression. But linear regression calculates an equation that allows us to make an "educated guess" at a person's score on some measure, given their scores on other measures. For instance, we might want to predict a person's GPA in college, given their GPA in high school. These variables are probably correlated (!), and information about one (high school GPA) gives us information about the other (college GPA).

- Z scores are standardized ways to report data. In fact, using the mean and standard deviation, we can convert raw data (the scores as given by the participants) into standard scores (or Z scores, same thing), by some easy arithmetic (bet you can hardly wait, right)? But standard scores are very useful, since they allow us to compare apples to apples, instead of apples to oranges.

In a normal ("bell shaped") distribution, the mean, median and mode will be smack in the middle of the curve, with 50% to the left and 50% to the right. The mean of a Z score distribution will be 0, and the standard deviation will be 1. We can use the Z table in the appendix of most statistics books to find various percentages between different Z scores, as well as percentile rankings for different raw scores.

For instance, if we have a normal distribution with a mean of 100 and a standard deviation of 15, then 1 SD above the mean is 115, or a Z score of 1.0. It also means that if we know the Z score, the mean, and the standard deviation, then we can find the original (raw) score. If you put the numbers for Z, the mean and the standard deviation into the formula on page 160, you can find the raw score, in this case 115. But you can find ANY raw score by putting in the Z score, and ANY Z score by putting in the raw score in the equation. Let's look a bit more at this . . .

The Standard Normal Distribution is a special one of the family of normal distributions; it has a mean of 0, and a standard deviation of 1. You can convert any normal distribution to the standard normal by finding the z scores of each raw score. Z scores (also called standard scores) tell you how close or how far away a score is from the mean. So for example, a Z score of 1.2 means that the score is 1.2 standard deviations above the mean, and a Z score of -2.1 means that the score is 2.1 standard deviations below the mean.

Now if we look at a Z table (which is always included in any statistics book), we see that between 0 and 1 we have about 34% of area. Combine this with the 50% of the area below the mean, and we see that a z score of 1 is at the (34 + 50) or 84th percentile. So about 84% of the people taking this test will score below 110, and about 16% will score higher.

Descriptive Statistics are used to organize and summarize a set of data. They can include the mean, median, range, standard deviation, percentage or other common calculations. The important thing to remember is that they ONLY apply to the specific set of data at hand; they are not intended to make judgments or inferences about other distributions, even ones that are similar. One extra important note to keep in mind: many categorical variables (such as gender, religious orientation, ethnicity, and so on) can only have descriptive statistics applied to them, in the sense that they can be counted. So we can say we have 20 males and 40 females, but we cannot say the "average" gender is (20 + 40)/2 is 30. Obviously this has no meaning! When you see the term alpha (usually .05) we are looking at the probability of making an error (Type one error). More on this soon, but the .05 refers to the TAILS of the Z distribution. For example, if we take 95% of the curve in the middle, this leaves 5% left over. If we split that 5% into the 2 tails, we have 2 1/2% in each extreme of the distribution. These are very far away from the mean, and are not likely if the mean is a true parameter.

Now, if we have a hypothesis that the actual mean of a sample is X, we can test that with the Z (or t) procedures; Here is the important part:

IF THE TEST STATISTIC IS IN THE TAIL OF THE DISTRIBUTION (THAT IS, IT IS NOT IN THE 95% MIDDLE PART), THEN WE HAVE A STATISTICALLY SIGNIFICANT RESULT.

The Z score of +/- 1.96 is a VERY important cut off point; literally, it is where we draw the line between retaining a null hypothesis or rejecting a null hypothesis. Notice that these two scores represent an alpha of 05! Also, you might see that 2 1.2% of the distribution is on the left, and 2 1/2% falls on the right. These are the rejection regions. Again, if the calculated Z score is in the rejection region, we have statistical significance (the result is PROBABLY not due to chance). Remember that this does not PROVE that the null is false; instead, it SUGGESTS that PROBABLY the null is not true).

Power, Type 1 and Type 2 errors

Statistical power, type 1 and type 2 errors are some of the more abstract concepts that students have to come to terms with. Let's take a look at what they mean!

Recall we have a null, and it is either true or false. We have an alternative, and it too is either true or false. We use our statistical analysis to determine the likelihood, or probability, that the null is true, and make a decision based on those results.

But we could be wrong in our decision!

A type 1 error occurs when the null is true but we decide to reject it. The probability of making a type 1 error is alpha (usually, .05 or .01).

A type 2 error occurs when the null is false but we do not reject it. This is called beta, and is related to statistical power.

Power is the probability of rejecting the null when in fact it is false. So power, in this sense, is a good thing, and the more power in an analysis, the better. Power is equal to the value of (1-beta).

You: *Ya know, I kinda sorta get it, but can you give me an example of this?*

Me: OK, how about this one (since I am getting hungry)!

Pretend you are making a big pot of chili, and you wonder if it has enough salt. You, being a great statistics student (and taking a great class from great instructors . . . hey, we need to get in on this too), frame the following null hypothesis:

Ho: The chili is properly seasoned (it does not need any additional salt).

The alternative, of course is

H1: The chili is NOT properly seasoned (it needs more salt).

Now, either the chili IS properly seasoned, or it isn't (if we knew, we would not have to make any hypotheses about it)! If it is seasoned properly, and we decide it is seasoned properly, we have made a correct decision. On the other hand, if it is not seasoned properly, and we decide it is not seasoned properly, we have also made a correct decision! This latter decision is called Power (rejecting the null when it is false).

On the other hand, if the null is true and we reject it, we have made a type I error (in our chili example, we conclude that it needs seasoning, when it does not). We end up with over seasoned chili! Your party guests will be displeased (see how useful this class is)?

But, what if we conclude the null is true (it is seasoned properly) but it is not (it really needs seasoning)? This is a type II error, and we end up with under seasoned chili! (Another major faux pas for you party planners)!

Now, you might be wondering about how all of this is related to the sample size in a study. After all, we hear all the time that larger samples are better, but why?

Let's consider this a moment; if you take a SMALL taste of the chili, it might need seasoning but your small sample missed this. In the same way, if your sample size is too small, there may be real differences between groups, but you will not be able to detect it (power is too low).

But if you take a larger sample of the chili, you are more likely to recognize it is under seasoned (that is, the null is false), and add some salt (that is, you reject the null). Viola! You have achieved statistical power, at least in the culinary sense! The parallel to a real study should be apparent.

The trick of course is choosing the correct sample size. Too small will miss it, and too much is wasteful. The same thing happens in research, and when YOU do your own research, you will have to justify the necessary sample size to achieve a given level of power. Several (free) programs to calculate power and sample sizes are available online, and many statistics books have tables that can also be used for this purpose.

YOU: *This is just incredible. I would have easily paid 5 times as much for this book as I did. But can we take a break from the math and stat stuff and talk a bit about writing? How do I present all of the results?*

ME: Glad you asked! Let's shift gears a bit and answer these important questions!

Chapter Four

It's Not Just What You Say, It's How You Say It

As you progress towards the attainment of your Bachelor's degree, or the creation of your Master's Thesis or Doctoral Dissertation, the faculty will be expecting more and more of you, including the quality of the writing that you present to us. It is a MUST that you speak, think and write like a professional in the field—your writing represents you. In your statistics course work, you and your instructor may well concentrate on the results of your analyses, and many of the assignments you will turn in are to be considered practice in proper presentation of results. However, it is not just doing the correct analysis properly, but presenting the results in a manner that is clear, organized, logical and professional. You will be expected to use proper terminology, and a tone (voice) that conveys a serious attitude towards the results. Instructors are interested of course in your ability to PERFORM the statistical procedures, but just as interested in your UNDERSTANDING of the output and your WRITE UP of the results that show your understanding.

YOU: *What?! I took a statistics course because I thought there wouldn't be any writing—who knew? How do I get started?*

ME: Start by understanding yourself as a writer. In general, there are two kinds of writers: outliners and drafters. Both are trying to get to the same place, but they come from different directions. Outliners begin with a lot of ideas and minimal text; drafters begin with a lot of text and find their ideas as they write. Therefore, outliners have to expand the text to explain their ideas fully, and drafters have to cut their text to reduce repetition and make their ideas more clear. So which one should you be? It's a personal preference. And it may change, depending on the writing task. Do what works for you in a given situation.

The best way to begin is to identify how you generate ideas: are you an outliner or a drafter? Do you start by listing ideas or do you just start writing? Both have to revise and edit text, and can even use the same tools; they just use different strategies to get to the same place. If you don't know whether you are a drafter or an outliner, try both methods on the next assignment and see which one works best. You might even find that you combine both, outlining and writing pages upon pages of text, handing off all of those lovely words to yourself for revision. Understanding who you are as a writer will help you become more efficient and can help reduce the stress of beginning a writing project.

YOU: *Okay, I know who I am as a writer—what's next?*

ME: Once you find yourself as a writer, it's important to understand your audience and your purpose for writing to them. Who are they? What do you want to tell them? What are their expectations?

For help with audience analysis, we thank the ancient Greeks, who identified three broad groupings of rhetorical approaches: pathos, ethos, and logos. It is very important to match the approach with the audience. While the names may be new, you are probably familiar with the approaches.

Pathos is called the emotional appeal. We see this used a lot in advertising. Think of television commercials, particularly political commercials, where emotion is rampant. They play on your emotions to get you to do something. For instance, a commercial in Washington State has been on the topic of where to locate the new county jail. The commercial shows a woman holding her baby, sitting on a front doorstep, saying "I don't want a jail in my backyard, do you?" She is tugging at your heartstrings, so that you won't locate a jail near her baby in the next election. She has not provided any evidence that having a jail near her home would be a problem.

She is offering her opinion, which works fine for emotional appeals and narrative writing; however, unsupported opinion does not have the power to persuade a scholarly audience looking for statistical significant findings. Pathos also opens up the door for an unintended reaction when posing a question. What if, in the example above, the audience said, "Sure, I'm happy to have the jail in my back yard; I'll be closer to work that way"? The argument loses steam immediately.

Use pathos rarely and carefully because not everyone will answer the question or relate emotionally to the issue with your intent in mind. When used in formal research documents, this approach will cause the scholarly audience to stray or lose interest. Worse yet, it could cause you, as author, to lose credibility with a scholarly audience.

Ethos is called the ethical appeal. This appeal establishes the credibility of the author, both you and any of the authors you choose to use in your research. When you evaluate your sources, you are looking for the ethos, the credibility and expertise, of the author of the source. When you use credible sources, you are working on your own ethos as a writer.

The ethical appeal is essential for establishing credibility. To aid in your search for credibility, you will be asked to use a citation and reference style, most likely APA (more on that later), which is the favorite style for social sciences. Any time you use an idea or the words from an outside source, you must cite that source to establish its lineage and credibility. Citations also help establish your credibility as a writer because they signify that you know the existing literature on the topic, while giving the original author credit for the idea. Use ethos liberally in a research paper!

Logos is called the logical appeal. As it sounds, this appeal uses logic and evidence to support a claim. Think of crime solvers when you think about how this works. Just like a crime scene investigator (CSI), you have to identify a problem and gather the facts and evidence in order to solve the problem. The only difference is that you don't have a dead body to deal with, but you do have a unique audience with particular tastes for evidence.

Just like a CSI, as you gather your evidence, you have to organize it in a way to make a sustained argument. While a CSI might organize around the crime—Col. Mustard did it in the library with a rope—your organization is going to depend on your main point. The CSIs have to provide evidence that it was Col. Mustard, just as you have to provide evidence that your main point is valid. The logical appeal is THE NUMBER ONE appeal that you should be using in research papers (mixed in with a lot of ethos, of course!).

YOU: *I've heard a lot about a "third person" in writing. Who is this and where do I find him or her for help?*

ME: Third person is a "what," not a "who," and it can definitely help with your writing. It describes the point of view of the author.

First person: I, me, my, we, us, our, ours

Second person: you, your.

Third person: He, she, it, they, them, his, hers, its, theirs

Finding the correct answers for any data analysis is obviously a high priority for any researcher. But, reporting those results, using a presentation style that has been endorsed by the psychological community, is equally important. APA uses precise language that is always in third person.

There are no "I, we, let's, our" or other personal pronouns. There is no "This researcher, this learner, this student" form of third person (it's actually a really clunky form of first person used in journalism, not scholarly texts). Always and only use proper presentation of third person in your write up. Remember that this is excellent practice for you as you get closer to a dissertation or thesis.

YOU: *So, then how do I write in third person?*

Writers are expected to write in first person, using "I, us, we, our," when they are writing personal narrative essays. Not only are the essays first-hand experience, they are also reflective in nature.

Research papers using statistical analysis, however, require third person, using "he, she, or it." "I" (first person) and "you" (second person) are not appropriate. "One" is acceptable, yet more formal in tone.

So how do we make the move from first person to third person? By banning the use of "I think," "I believe," "in my opinion," "I agree," and other statements like this that insert you, the author, into the text. These phrases are considered unnecessary and redundant because the reader knows that you are the author; therefore, everything in the document is your thought, belief, or opinion, which you are substantiating with evidence from the research. It is the nature of academic writing: You no longer have to announce yourself in your text (which is, actually, a journalistic convention, not an academic one).

YOU: *You know by now I am going to ask for an example.*

ME: Of course! For instance, if I said:

In my opinion, the sky is falling.

It is obvious that the thought is mine because I am the author; it can be distracting to the reader for you to call attention to yourself in an academic text. What if I said instead:

The sky is falling.

The sentence has the same content and it's obvious that the thought is mine since my name is on the paper—but it is still just an opinion. In academic persuasion, we don't want opinion; we want informed opinion, evidenced opinion, which means making a claim, providing evidence, and adding reasons and analysis. So, what I really need to say is this:

The sky is falling. According to Smith (2015), the Earth's gravity has gained ground and is pulling the stars and moon closer to the Earth's surface. This gravity increase has created a "sky is falling" effect rather than an "Earth is pulling" effect. Scientists are looking into what many are calling an optical illusion.

This, in a nutshell, is the essence of academic research writing. The first sentence is the claim or main point. The second sentence is the evidence, cited in APA style. The third sentence is my analysis or critical thinking on the subject. Notice how the third sentence ties together the main point and the evidence. The evidence says only that the gravity of the Earth has increased; my analysis says that it is creating a "sky is falling" effect.

YOU: *Since I'm only reporting on statistics, do I need a thesis or a main point?*

ME: YES! You need a thesis and a main point. You are not simply reporting the statistics. You are reporting on the statistical analysis, but you are doing more than a report. You are creating new knowledge by telling the reader what the statistical analysis means from your perspective. Here are a couple of ways to look at the difference between an informational report that provides, well, information, and an analytical report that provides information and an explanation of what all the information really means. You should be aiming for the analytical report.

The battle royale between writing a report (sponsored by "just the facts") and writing statistical analysis with a thesis (sponsored by "I have ideas too") will be held this quarter—make sure you know which side you are on!

In no particular corner, representing the lightweight division, Report tells us what SOMEONE ELSE thinks and gives us just the facts when weighing in on the topic. You must cite the sources in the report because they are not your ideas. In the same location because both genres are valuable, representing the heavyweight division, Analysis gives us YOUR thoughts when weighing in on the topic after reviewing and making conclusions from your statistics. YOU are also the referee for the bout, which means you have to make some kind of Judgment, also known as your Position, which is part of your Thesis. When you make that judgment, you must present it proudly like a boxer holding up gloved hands in victory for all to see: your Thesis is what fuels the Argument and provides organization and logic to the paper.

In other words, Reports are the building blocks of statistical analysis. They provide information and only information. This information comes from an outside source and represents other people's ideas. You ALWAYS cite other people's ideas. Analysis is critical thinking. You look at the facts and evidence, and then break them apart to examine them, developing your own ideas on the topic based on the facts and outside sources used as evidence to support your insightful thinking. You do not cite your ideas—that's how the reader knows they are yours.

Judgments or recommendations are the conclusions you come to after your analysis of the information. These conclusions make up your Thesis, which is presented in your analytical report. The Thesis is what fuels the Argument and provides organization and logic to the paper.

YOU: *You know the next question: How do I write a compelling thesis statement?*

ME: Thought you'd never ask! A thesis statement provides the reader with your purpose for writing and usually comes at the beginning or the end of the introduction to your paper, which provides the background and context for your purpose.

Here is a model to use as a check for the essential elements. Generally, a thesis statement comes in two parts: 1) the claim and 2) the reasons for the claim. Here is a model using these two parts:

This is so . . . because . . .

"This is so" is the claim. You are saying that something is valid (not the Truth, but valid).

Chocolate should not be banned from diets . . .

"Because" is where you put the reason(s) for the claim. These are the reasons that led you directly to your conclusion, which is, yes, your claim.

. . . because new health benefits have been found for people who eat chocolate in moderation.

Putting these two together gives you a thesis statement:

> Chocolate should not be banned from diets because new health benefits have been found for people who eat chocolate in moderation.

To add strength to your thesis and provide organization to your paper, you could add what we call a road map, which provides a preview of the evidence and an organizational pattern.

Chocolate should not be banned from diets because new health benefits have been found for people who eat chocolate in moderation, specifically in the areas of Alzheimer's prevention and depression management.

From this thesis statement and road map, I would expect the introduction to explain the problem and provide some background (chocolate is generally considered empty calories and a cause of weight gain, diabetes, and other health problems, which is why it gets banned). The thesis will appear somewhere in the introduction.

Then, the body of the paper would:

1. Explain that chocolate has been given a bad rap all of this time and actually has been found to have health benefits.
2. Discuss health benefit # 1—dark chocolate in moderation has healthy antioxidants, using evidence to support this claim from the sources that you research, cited in APA format.
3. Discuss health benefit #2—chocolate has been found to help with Alzheimer's prevention, using evidence to support this claim from the sources that you research, cited in APA format.
4. Discuss health benefit #3—chocolate has been found to help with depression management, using evidence to support this claim from the sources that you research, cited in APA format.

The conclusion would then summarize the main points in the introduction and body, and then reiterate the thesis. Now, I'm not saying that every thesis statement has to match this form; however, it is a good place to start when clarifying your thesis and setting up the organization of your paper.

YOU: *I think I'm getting it. I don't want to look like I'm giving only my opinion, so how do I integrate the outside sources into my paper to support my claims?*

ME: The best tool I have found so far is the MEAL Plan (thanks to the Duke University Writing Center web site,https://twp.duke.edu/uploads/assets/meal_plan.pdf, for this jewel):

The MEAL Plan is an essential tool for writing with research. In a research paper, your tone should be objective and formal, yet confident and persuasive. You are, ultimately, arguing a position, and you want people to listen. The biggest challenge in a paper like this is to keep the emotion and opinion out of your writing (emotions and opinion in research writing work like reader repellent!). You do this by staying in the third person, making claims, providing supporting evidence, and building an argument through analysis. The MEAL Plan can help you with all of this. Here is how it works:

M. The sky is falling. E. According to Smith (2002), the Earth's gravity has gained ground and is pulling the stars and moon closer to the Earth's surface. A. This gravity increase has created a "sky is falling" effect rather than an "Earth is pulling" effect. L. Scientists are searching for the right tools to reverse gravity and ultimately, the falling effect.

M. This is the claim, main point, or topic sentence of the paragraph. It usually comes first, but can be found elsewhere in a paragraph, depending upon the placement of other elements. Don't fall into the trap of asking your audience a question that they may answer differently than you intend. Making a claim means making a statement, not asking a question (for instance, Is the sky really falling?). Questions are great to lead you through your research, but they should be answered, not just asked, in a final draft of a research paper.

E. This is the evidence that supports the claim that the sky is falling. Notice it has been taken from an outside source, and thus needs to be cited (this citation is APA style). Notice that evidence does not stand on its own—don't let the evidence speak for you—you tell the audience how the evidence relates to your claim. You do this with your analysis.

A. This is the analysis—your thoughts on the situation. Notice it is not cited because it is your idea. Your idea rules in a research paper. You are using research to find a position based on the evidence and argue for your position by showing how the evidence logically leads to your conclusion. This is how your opinion becomes "informed" opinion, an idea substantiated by the current knowledge yet adding something new to the knowledge pool.

L. This is the link or transition to the next paragraph which should be about the scientists' search for tools. As you continue to use transitions from one idea to another, you are building an argument based on your analysis of the evidence.

Now, I am not saying that all paragraphs are four sentences long or in this order. They can be in different orders or even not there at all—though it is a rare occasion that a paragraph does not have a main idea. On the contrary, I am saying that these four elements are usually in a paragraph and to make sure that your paragraphs have all of the information needed to make sense to your reader. The MEAL Plan serves as a model for writing with research, so begin with it, experiment with it, and grow with it.

YOU: *I noticed that you used and cited other peoples' ideas in your MEAL Plan paragraph. How do you decide whether to use their words exactly or to paraphrase what they said?*

ME: Terrific question! Use direct quotes (the other author's exact wording) sparingly, and only when a direct quote will create impact. You should be paraphrasing most of the time because you are the author of your paper, and your voice should dominate the paper. If you string together a lot of direct quotes, you will create a patchwork quilt of other people's work, and your voice, your argument, and your conclusions will be lost in the patches.

YOU: *And how do I paraphrase?*

ME: Another terrific question! Let me back up a bit and talk a little about paragraphing first. As you have seen with the MEAL Plan, paragraphs are made up of different elements: main idea, evidence, analysis, and link. All paragraphs will have a main idea; however, some paragraphs may simply provide evidence or analysis, some will be longer than others, and in others the order of the elements may change. For instance, I can certainly see a paragraph in the LEAM form, leading with a link from the previous paragraph, moving into the evidence, then analysis, ending with a main point (this is also known as the indirect approach). In longer documents, like books, a link may even be a paragraph long. If your paragraph is really long (rule of thumb: you should have at least one paragraph break per page) or really short (you know what that looks like :-), then look to the MEAL Plan to help you divide paragraphs logically and develop them fully.

NOTE: How long should a paragraph be? As long as it takes to cover one idea. One idea, one paragraph. When you find yourself wondering if you need a paragraph break, think about the elements of the MEAL Plan. Have you developed the paragraph completely using the elements as a baseline? Do the evidence and analysis relate directly to the main idea (also known as a topic sentence in paragraphs)?

Paraphrasing, especially in academic texts, is essential to paragraphing. It means combining other peoples' ideas with your own critical thinking to create new knowledge—the ultimate purpose of academia. You, as author, are expected to use your voice to inform the reader of the existing knowledge by using and citing outside sources. However, trying to explain someone else's idea in your own work can be tough, particularly when you just can't think of a way to explain it without directly quoting the author. A technique I use is to read the passage, turn it over so I cannot see it, and then write my interpretation of the author's intent using my own words. I've found that I have more success when I imagine I am explaining the idea to my best friend and don't have the text to review. But even that is not always enough to get the paraphrase just right.

Paraphrasing your own paraphrase of someone else's idea—paraphrasing squared—is a technique used to make sure you are not unintentionally plagiarizing material by using too much of the original language. It also benefits "voice." When you paraphrase twice, the material is more likely to sound like it came from your voice rather than someone else's voice, which makes the language flow like a fast stream. When there are too many voices in the text, it sounds like that patchwork quilt (hmmm . . . what does that "sound" like? . . . like a bunch of things that don't really go together but are being stitched together anyway).

Here is how you do it.

1. Read the section that you want to paraphrase
2. Turn the page over and don't look at it again (yet)
3. Write down what you understand the section to say
4. Reread the section in the article, turn it over and then reread and revise your first try at the paraphrase.
5. Make sure that your paraphrase is true to the original author's intent, but in your words. If you haven't met these criteria, repeat steps 1-3. If you have, go to the next step.
6. Cite the source using APA style . . . paraphrases must be cited because the ideas still come from the source.

It takes a little practice, but will save a lot of time agonizing over wordsmithing, trying to find a different way to say something.

YOU: *You promised more information on APA style citations and references. What do I need to know?*

ME: Ah yes, APA. References and citations work together to help you give credit to the original author, support your views with experts in the field, and help you avoid plagiarism. In brief, in-text citations lead the reader to the source on the reference page, just in case the reader wants to review the original document. For example, if the author of a 2013 document is Smith, then the in-text citation would look like this (just one of many examples, as you will see below):

Smith (2013) reported that (note the use of past tense)

And the matching reference would look like this, making it easy for the reader to find the publication information:

Smith, A. (2013). Name of the article. Name of Journal or Web Site. Retrieved fromhttp://url.goes.here.

Note that the second and all subsequent lines are indented (called a hanging indent). Both single and double spacing are acceptable for reference page entries. If you choose to single-space, then double-space between entries.

I have compiled some examples for your convenience, but know that you can always find more information and examples of APA citations and references at the Purdue OWL (http://owl.english.purdue.edu/owl/resource/560/01/). Each example below begins with a poorly crafted in-text citation and ends with possible revisions. Yes, that's right. There is more than one way to cite a source.

Original (common in-text citation):

The article that I read was "When IT is asked to spy" by Janet Marshall. The article talked about how many companies are asking their IT departments to monitor employees and their use of internet, e-mail, and cellphones.

Revised:

Marshall (2010) stated that many companies are asking their IT departments to monitor employees and their use of Internet, e-mail, and cellphones.

OR

Many companies are asking their IT departments to monitor employees and their use of Internet, e-mail, and cellphones (Marshall, 2010).

OR

According to Marshall (2010), many companies are asking their IT departments to monitor employees and their use of Internet, e-mail, and cellphones.

Note: use only the author's last name and the year of publication in parentheses. You don't need the name of the article in the text—instead, the article name goes on the reference list. Also note the use of past tense when citing other sources. Pay particular attention to the punctuation and spacing.

Original (attribution and citation):

"Tone in Business Writing" is a handout by Maya Kennedy and is provided on the OWL website. It begins by providing a definition for tone. Tone in a sense is basically just another way of referring to the attitude of the writer. Just like in verbal communication the tone is what can determine how a message is perceived. In verbal communication a person would choose his or her tone based upon the type of conversation. The same is true for writing.

Revised:

"Tone in Business Writing" is a handout by Maya Kennedy (2010) on the OWL website, which begins by providing a definition for tone. Tone is another way of referring to the attitude of the writer. Just like in verbal communication, the tone is what can determine how a message is perceived.

In verbal communication a person would choose his or her tone based upon the type of conversation. The same is true for writing (Kennedy, 2010).

Note: the first time the author is named, it is okay to use the full name (but you don't have to), but after that you should only use the author's last name and date of publication.

More importantly, the attribution at the beginning of the paragraph (Maya Kennedy, 2010) and the citation at the end of the paragraph (Kennedy, 2010), signal to the reader that all of the information in between the two comes from the same source: Kennedy. This is important to remember as a tool to avoid plagiarism.

Original (direct quotes):

"And since so many executives imagine themselves able to judge good writing strictly by intuition, we have grown contemptuous and cavalier about language, about reading, about writing, about the editing or writing, and about communication," stated Meyers (Meyers, 1961, p. 109).

Revised:

Meyers (1961) stated, "And since so many executives imagine themselves able to judge good writing strictly by intuition, we have grown contemptuous and cavalier about language, about reading, about writing, about the editing or writing, and about communication" (p. 109).

OR another possibility:

Research has shown that "since so many executives imagine themselves able to judge good writing strictly by intuition, we have grown contemptuous and cavalier about language, about reading, about writing, about the editing or writing, and about communication" (Meyers,1961, p. 109).

OR, if the direct quote contains 40 or more words, you need a block quote:

Research has shown that since so many executives imagine themselves able to judge good writing strictly by intuition, we have grown contemptuous and cavalier about language, about reading, about writing, about the editing or writing, and about communication. Just pretend that I added more words to make the 40 word limit, so I could show you the format. (Meyers, 1961, p. 109)

Note the block indent, the lack of quotation marks, and the only time that an in-text citation will come after the period.

YOU: *I know that writing up results is really important, but I am not sure how to do this. Can you give me a couple of examples?*

ME: Sure I can; here is a poor example of a hypothetical write up:

"We used an ANOVA to see if there were any interesting results. We looked at the independent variables of gender and age, and the dependent variable was reaction time. We found that there were some differences between males and females, but not between different ages. Also, there was no interaction at all."

Now, compare that to this:

A two factor analysis of variance (ANOVA) was performed with independent variables of gender and age, and a continuous, quantitative variable of reaction time as the dependent measure. Main effects for gender were found to be significant, with reaction time for males (M = 2.3, SD = .9) significantly higher than that for females (M = 1.8, SD = .88). No significant differences for age were noted, nor was there any significant interaction effect for gender and age.

Get the idea? What specific difference can you see in these two examples? They both convey information, but HOW the results are expressed distinguishes the proper one from the clumsy one. Your APA 6th edition manual has specific pages that you should consult. In addition, many of the newer statistics text books include a sample write up section at the end of each major chapter, and would be an excellent format for you to use in your write ups!

YOU: *This is really good stuff. Man, I feel like I should have spent a LOT more than I did for this book. Can you give me your address so I can send some additional money?*

ME: I'm flattered you feel that way! Tell you what...let me give you some additional tips and comments, and we will call it even. Here are some general comments that I hope you will find helpful when it comes time to put pen to paper. Some of these comments apply to your thesis or dissertation proper, but many also apply to the assignments you will be writing and turning in to your instructor:

1) Plan to spend about twice the time you estimate it will take to do the entire assignment. In other words, do not underestimate the time you will have to allocate to properly research and write up a major paper. We (that is, the faculty) well understand how busy you are with careers, family, and social obligations. We applaud your determination and dedication to pursue an advanced degree. However, be this as it may, these kinds of major documents involve a considerable amount of time on your part, and should be part of your overall plan. In addition, your friends and family should be aware that you may not be as available during the planning, execution and writing of a thesis or dissertation.

2) Quality counts. Never send your professor anything less than your very best work. Even if it is an email, a draft, or a question, use clear, clean and professional communication. Your expectations as graduate students are much higher than those for undergraduate students, and I guarantee your professor or mentor will be disappointed if you turn in work with spelling

errors, punctuation mistakes, poor organization, hasty, superficial commentary, missing tables, and other such writing blunders. If something is taking more time than expected, write a polite note and ask for a modest extension. Many will be happy to grant this to you, and all will appreciate a better work.

3) Be prepared to redo your work, especially at the Masters and Doctoral level. Thesis and dissertations are iterative, meaning that what you submit will most likely be modified, corrected, or in some other way changed. Our intention here is not to burden you with extra work, or to make you feel inadequate. The honest truth is that you will have a much better manuscript if you follow the advice and recommendations of an expert, in this case your mentor or professor. He or she will be eventually signing off on your work, and I can guarantee they will not do so until they consider it as near to perfect as possible. So please take comments for revision in the right way; it is not a good idea to "cop an attitude" or engage in an argument. On the other hand, it is perfectly fine to explain why you presented something as you did! Mentors may mis-read or mis-interpret something, and a respectful clarification is perfectly appropriate.

4) Many Masters and Doctoral students will have a mentor, a senior faculty member who will guide your work. Your mentor has more students than you probably realize. He or she may have between 5 and 10 people that they are working with at the same time. In addition, many adjuncts teach at more than one University, and also have 5 to 10 students at those locations. This does not count the 1 to 3 classes they have been assigned to teach, most with about 20 people in each one. The point is, you may have to wait for a response. It is wise to try to submit major sections or drafts early, expecting that it may take your mentor one to two weeks to read and comment on it. Also, if you submit something over the break, a mentor may not respond, as faculty need some down time as well. During the term, if more than 2 weeks have gone by and you have not heard back from your mentor, a polite reminder note is in order, perhaps with a re-attachment of the work.

Your writing is really the single most important aspect of your education. Perhaps that sounds like an exaggeration (if so, good critical thinking!), but we must rely on our written (and sometimes oral) communication to understand and get a "feel" for the individual.

How do they express themselves? Are they clear and articulate, or are they overly wordy? Do they have flawless spelling, punctuation, grammar and sentence mechanics, or are there errors in these basics? Are they following established templates when provided, have good organization and transitions throughout the paper, and end with strong, logical conclusions? Is the paper free of opinion and obvious bias (as a research paper should be), or are the author's wishes seeping through? Do they use the proper vocabulary, pertinent to their profession, or are casual, common words creeping in? Are they consistent in their terms and definitions? Have they proofread everything to make sure it is as perfect as possible, or had someone else do this for them?

These are just some of the things you will find as you continue to learn, research and write professional papers. We have seen that YOU are responsible for critiquing the works of others, so it is no surprise that your professor and eventually your committee chair (or mentor) will do the same with your submissions. Always put forth your best efforts; remember that first impressions do count! Here is a slightly modified example of two student letters sent to me requesting my services as their mentor:

Student 1 writes: Dear Mr. Trunk, Hello, my name is Paris Hilton, and I am in need of a person to help me with my research. I am kinda sorta interested in the relationship between wealth and safe driving habits. I guess you are pretty good at math and statistics, because I saw that you teach this stuff. So I was hoping you could assist me in my work. Thanks a lot, and please let me know.

Student 2 writes: Dear Dr. Trunk, My name is Britney Spears, and I was a student in your advanced research course last Spring. I am currently organizing my dissertation committee, and would appreciate your consideration in possibly joining as a methodologist. My proposed topic is quality parenting in the entertainment industry, something I hold close to my heart. Thank you for your time and consideration of my request; I have sketched my preliminary prospectus, which I would be happy to send to you if you wish to see it. Thank you again Dr. Trunk.

Obviously these are a little doctored, but these are not too far off from the kinds of requests we as faculty receive almost weekly. If YOU had to pick one, which one would you choose? I guarantee if your writing is clear, succinct and professional, you are way ahead of the game in finding an available chair down the road.

YOU: Thank you! I never really thought about statistics being much more than formulas, number crunching, graphs and charts. But now I realize that the writing, presentation, and display of the information, especially as it demonstrates a professional tone and voice, is very important!

ME: Thank YOU! My work here is done (but I may have more for you later)!

Chapter Five

Some Common Issues within the Field of Statistics

YEAH, YOU'VE GOT TO KNOW THESE

Hey! There you are! Great to see that you are still with us as we continue our journey into navigation and success in statistics! If you have come this far, you realize that there is much more to being successful in a course in statistics than just getting the "right answer" on a homework or exercise problem! There are definite, well-defined strategies for studying, analyzing, and writing your assignments that will not only help you to achieve higher marks in the class, but (perhaps more importantly) help you to really understand what statistics is all about.

Statistics is really a way of thinking about the world, often in terms of probability, one of the topics we will be covering in this chapter. Additionally, in a real research project, such as a capstone assignment, a Master's thesis, or a doctoral dissertation, other important issues must be addressed. For example, the size of the sample, as you know, is important. So is how we recruit participants (we no longer use the word "subjects" when referring to human participants, but this is an acceptable term for animal studies), how we select participants for inclusion in the study, how we place them into groups (randomly or non-randomly), how we prepare for attrition (the loss of participants over the course of the study), and how we present all of this information in written form, either for our professor or for a professional journal editor.

Other issues concern the data themselves. How do we know that the data are valid? That is, how do we know that responses participants give are true indicators of their inner character or personality? Put more bluntly, how do we know we are not being lied to? After all, the only thing we have in front of us are a bunch of numbers, and while some people may provide numbers that are true and meaningful, others may not be so kind, and give us numbers that are meaningless! For instance, some people might want to get the task "over with" and simply mark all 6's for every response. Others may not understand the instructions, may not be fluent in English, or may deliberately lie in their responses! So the general question of data that is valid, and data that is reliable should be carefully considered, not only in a real research study, but as part of your critical thinking strategy when reading an article, or looking over an example homework problem. Keep in mind that all the fancy statistics, tables and graphs mean nothing if the design is flawed, if assumptions are not met, or if a sufficient sample size is not obtained.

Certainly related to this is the issue of outliers, which are scores that seem to be very different from the typical ones submitted. What do we do with them? Can we just toss them out, sweep them under the rug, and pretend that they never even existed? Can we change them to be more in compliance with other data we have gathered? Can we analyze the data twice, once WITH the outliers and once WITHOUT them, and see if there are any real differences? Ah . . . so many possibilities . . .

Another other important issue has to do with something that seems to come up a lot in data collection, and that is the controversy between ordinal (often reported as Likert style data) being used in parametric statistical analyses. I know this is a bit on the geeky side of things, but it is a fairly common occurrence, and we should know what the proponents on both sides of the argument have to say, so that we ourselves can make an informed decision on this topic.

And since we are on the topic of ordinal scales, we should get into the closely related topic of the typical assumptions for many of the common parametric procedures you are likely to study. As we will learn, it is not only important to be aware of these requirements for valid statistical analysis, but often necessary to check that they are met long before the data are run!

Finally, you may hear from your instructor, or read in your text book, that sometimes data can be "transformed," and you may, quite rightly, ask what that is and why it is done. Good question! We will definitely have to talk about this in the chapter!

OK! Let's get into it, shall we?

YOU: *I know that probability is really one of the fundamental cornerstones of statistics. I recall your really eloquent and descriptive explanations of alpha (probability of a type 1 error), and beta (probability of a type 2 error) in my earlier readings. But can you talk a bit more about probability? And is this the same thing as odds?*

ME: You know what is really interesting about probability (OK! I see you back there! Yeah you, the guy thinking "nothing at all)!

Actually, what is really interesting about probability is that we use it all the time. I mean every day, almost constantly. The only thing is, we usually just don't recognize that we are doing so. Let's look at some common examples, OK? How often have you done, or thought about, some of these things;

If I leave for work now, I can (probably) miss most of the morning traffic.
Hard work usually (probably) results in better scores and higher grades.
Saving a little money from each pay check will (probably) help later on in my retirement.
That new dog (probably) needs a lot of supervision to house break him.
If I buy a lottery ticket, I will (probably) lose my money.
If I practice for my job interview, I will (probably) make fewer mistakes and be more comfortable.
If I am asked to think of a card, and to then pick a card from a deck at random, I have a small chance of choosing the card itself, a better chance of getting the suit I am thinking of, and an even better chance of getting the color (black or red) of the card I am thinking of.

You can (probably) think of many other examples, right? (and if you come up with some really good ones, email us and let us know what they are)! But of course the common element here is the notion of chance, of uncertainty in the outcome, and some kind of a system that quantifies the degree of uncertainty. Let's look at the last example (number 7) in a little more detail to see what this means.

As you (probably!) know, a standard deck of playing cards has 52 separate cards, 26 black ones, 26 red ones, with four separate suits (clubs, diamonds, hearts and spades), of which there are 13 of each (that is, 13 hearts, 13 spades, etc.).

Now in example 7, you are asked to think of a card. Let's say you think of the 10 of diamonds. What is the probability that you could then pick a card at random, and, lo and behold, it is the 10 of diamonds?

If you think about this a bit, you realize that since there are 52 possible cards you can choose, and only one of them is the 10 of diamonds (a "successful" choice), the probability of doing this is simply 1/52. Another way of saying this is that if you were given 52 chances at this (say, having 52 decks), you would, on the average, be successful in picking the 10 of diamonds once in those 52 times.

Now, let's think about the second option here. What is the probability that you will be correct in thinking of a particular card, and then choosing a card of the same suit? So in addition to the 10 of diamonds, all of the other diamonds in the deck (2, 3, 4, up to the Queen, King and Ace) would count as a success. A little thought should convince you that the probability of this event is 13/52, or reducing this quotient, 1 out of 4. That is, since we have 4 equally sized suits of cards, we have a one out of four chance of having our card be the suit we thought of (diamonds, in this case). Using common language, we might say we have a 25% chance of choosing the right suit.

Get the idea? What would be the probability then of having the card you choose, after having thought of it, be the same color? We know that half the cards are red (the 13 diamonds and the 13 hearts), and half are black (the 13 clubs and 13 spades). Since there are 26 successes here out of 52 possibilities, we have a 26/52 probability of this outcome, or reducing terms, 1/2. Using common language, we might say that we have a 50% chance of choosing the correct color, or equivalently, a 50-50 chance of choosing a red card.

We see then that one way to look at probability with a finite set of possibilities is to calculate the ratio of the successes to the total number of possibilities. Now that wasn't so bad, was it?

One final thing that we should mention is the relationship between the probability of an event, and the odds of the event. In everyday, casual language these are usually used interchangeably, but they are not the same thing (although they are related, and in fact, if you know the value of one, you can find the other)!

As we have just seen, we can find the probability of a finite event if we know two things, the TOTAL number of possible outcomes, and the number of SUCCESSFUL outcomes. We saw how a few examples of this using cards, and we encourage you to make up some additional examples, just for practice, and (especially) to read the sections in your statistics book dealing with these concepts.

But before we go, let's talk about odds. In terms of a definition, we might say the odds of an event is the ratio (quotient) of the number of successes to the number of non-successes, or failures. Why don't we revisit the same examples as before, but this time calculate the odds of the events, rather than the probabilities.

Recall that in our first example, we found that the probability of choosing a card at random from a standard 52-card deck that matches a card we thought of was 1/52. If we took a calculator and divided 1 by 52, we would get the probability equal to .0190.

But now let's find the odds of choosing a card at random from a standard 52 card deck that matches a card we thought of. Using our formula, we take the number of successes (one, as there is only one card we thought of), divided by 51 (the number of non-successes). Our ratio becomes 1/51, or in decimal form, .0196.

While these at first glance appear to be very similar, the former represents the probability of the event, using the total number of possibilities in the denominator, and the latter represents the odds of the event, with the number of non-successes in the denominator.

Our second example, you may recall, found that the odds of choosing a card that was the same suit as the card we thought of was 1/4, or 25%. But what are the odds of choosing the suit? Once again, we substitute our numbers into the formula. There are 13 successes, that is, cards of the same suit (diamonds in our example), and 52 – 13, or 39 non-successes. So the odds of this event are 13/39, or 1/3, and in decimal form, 0.333. Notice that this last figure is NOT a percentage (only probabilities are percentages). Instead, it is the odds of choosing the thought of card.

Put another way, we can say that the odds of choosing the right suit are 1 to 3, and the odds against it are 3 to 1. That is, we have three times the chance of NOT choosing the thought of card as we do of choosing it!

One last example! Our final probability was in choosing a card of the same color as the thought of card, and found that this was 50%. To find the odds, once again, divide the number of successes (26, as there are 26 red cards) by the number of non-successes (also 26, as there are 26 black cards). Of course, 26/26 reduces to 1/1, or in words, 1 to 1. This simply means that we have an equal chance of selecting a red card as we do a black card.

As you can see, the odds of an event are closely related to the probability of an event, and in fact, there is a simple formula you can use to convert from one to the other. Here is how it goes:

IF YOU KNOW THE ODDS OF AN EVENT (which we can symbolize as X to Y), then the probability of the event is X/(X+Y). In this formula, as before, X is the number of successes, and Y is the number of failures.

To see this in action, recall that we just say that the odds of choosing a black or a red card are 1 to 1. So the probability of choosing the correct color card is just 1/(1+1) or 1/2, or 50%. In other words, if the odds are 1 to 1, the probability of success is 50%.

IF YOU KNOW THE PROBABILITY OF AN EVENT, THE ODDS ARE EQUAL TO X/(X+Y), where X is the number of successes, and Y is the number of non-successes (which is also sometimes called "failures" in some books).

We saw earlier that the odds of choosing the correct suit are 1 to 3 against being successful. There is one success (diamonds, you recall), and 3 non-successes (choosing a spade, a heart or a club would qualify as a failure), so that the probability of this event, using the above formula is 1/ (3+1) or 1/4, or 25%. We see then that odds of 1 to 3 are the same thing as a probability of 25% (remembering that probabilities are fine to express as percentages, but not odds)!

So there you have a brief introduction to probability and odds! How about we have more fun now (of course, my definition of "fun" may be different from yours) and move on to our next issue within the field of statistics.

YOU: *Thank you! This really helps. But you said something about sample size, sample recruitment, and sampling methods. Can you help me to better understand some of these issues?*

ME: I am so glad you asked! When people think of statistics, they think of numbers, formulas, calculations, tables, graphs and charts. But normally they do not think about HOW those numbers, tables and graphs were obtained in the first place. Where did they come from? How do we know they are accurate, valid and reliable (another good question, that we look at next)!

Data come from samples, which are portions of the population under consideration by the researcher. Now the word population used in statistics is not the same thing as the word population in casual conversation. Actually, a population is any closed, finite set under consideration by the researcher. He or she can define the population as needed. Here are a few examples:

All of the computer chairs produced by the ABC Chair Company from June 1, 2014 to June 1, 2015.
All of the students in Mr. Kent's high school math class.
African American women between the ages of 21 and 51 that have graduated from college.
Current active players in the NFL roster.
Democrats in Woodbury County, Iowa
All red cards in a standard deck of 52.

So clearly, the population can be ANY well-defined set of people, objects, or events that the researcher feels is worthy of study. Now, while a population may be of any size (the number of Democrats in Woodbury County is probably in the thousands), usually the researcher can only study a fraction of the available members of the population. There are many reasons for this, time and money not the least of them.

But interestingly, even a small portion of well chosen "units" from the population can give valuable information about the entire population. For example, if the ABC Chair Company produced 100,000 chairs during the year, we might be able to take a sample of 10,000 and inspect them for defects. If we find that 50% of the sample had a defect, we may be able to correctly infer that 50% of the entire population of chairs had a defect as well! Certainly the ABC Chair Company would want to know about that!

There are several things to notice about this last paragraph. One thing that comes to mind is that the larger the sample chosen, the more confidence we have that the conclusions we draw are valid. In other words, larger samples lead us to more valid conclusions when we make an inference from the sample to the population. How might your conclusions have differed if we had sampled 10 chairs instead of 10,000?

Secondly, the manner in which the chairs were chosen for inspection should be made clear to a reader. Did they take the first 10,000 off the line? Were the 10,000 all made in July? Were the chairs randomly selected according to some kind of random number schedule? The point is that HOW the chairs were selected is just as important as the NUMBER of chairs in determining the validity of the conclusions.

Third, notice that we are confined in our conclusions to other members of the population that might have been selected. If for example the ABC Chair Company also made recliners, but we only sampled computer chairs, our conclusions would be limited to computer chairs, and not to recliners or other kinds of chairs not in the sample.

Sample size is critically important in science, and especially in statistical analysis where information from the sample might be used to make valid inferences about the population from which the sample came from. As just mentioned, samples that are too small are suspicious in that we do not have sufficient information to make conclusions, and samples that are too large are redundant, time consuming, expensive and unnecessary. So, how do we find the sample size that is just right for our research question?

While it might be tempting to just pick a convenient number (there are 75 people at work I could sample, so my sample size will be 75), this is not recommended. Instead, there are three main ways that someone using statistics might go about this, and all or the choices depend on a few things regarding the data. We have to get a bit more technical here, so put on your thinking caps (as my dad used to say), and let's see what it is all about.

Sample size in general depends on three things: The first one is the level of alpha chosen (alpha is the probability of concluding that there IS something happening in the population, usually due to some kind of treatment given to one sample versus no treatment given to another part, when in fact there is not). In terms of hypotheses, alpha is the probability of incorrectly rejecting the null hypothesis of no differences. This incorrect rejection is usually called a "Type 1 Error" in statistics. Secondly, the level of power needed (Power is the probability of correctly rejecting the null when the null is false), and thirdly, the effect size (the presumed differences between the mean under the null and the mean under the alternative). In many statistical analyses, standard values are often assigned. For alpha, a value of .05 or .01, for power, a value of .80, and for effect size, a medium size.

Once these three values (alpha, power, and effect size) have been determined, we can either use a statistical software program to calculate the needed sample size, or we can use a table (in some, but not all, statistics books), or we can use a "rule of thumb" for certain types of statistical analysis. The main point to take away is that the sample size is determined by aspects surrounding the data and the data analysis, and not just a round number like 100.

YOU: *This is excellent information to know! I was one of those students who really didn't think much about sample size at all, because I was so worried about the math and calculating values for the data. Great to know this. It seems that statistics is really a lot more than just number crunching! Have you got some more for me?*

ME: I do! Let's talk next about HOW the sample is obtained. The first thing we usually think about, especially when we are working with human participants as our sample, is the form of recruitment. How do we find people that are suitable participants in the first place? Usually, we have to be pretty far reaching here. More conventional methods might be an ad in a newspaper or trade publication, a flyer on a bulletin board, a TV ad, or word of mouth. While 50 years ago these might have been acceptable, our modern digital age, combined with the widespread use of social media, makes these look rather archaic!

Many students and professional researchers now use the Internet to advertise their study, provide preliminary information about the kind of people that they are looking for (for instance, men with diabetes, or children in 3rd grade with ADHD), how long the participation might last, if there is any pay or incentive involved, and so on. Social media sites like Facebook and Linkedin can reach literally millions of potential participants, and if the questions or assessments can be delivered electronically, this is often the method of choice for recruitment. Of course, some studies require live participation in real time at a local facility, and in those cases a more traditional method might be advisable.

Once a potential participant has agreed to learn more about the study, additional information is provided so he or she can make an informed consent (if over the legal age of 18), or the parents of a minor can read and ask questions about what will be expected of their child if the participant is under the age of 18. Notice that parents give consent, while younger participants can assent to be in the study. Keep in mind that by far the majority of studies conducted with humans recognize the participants' right to terminate their involvement in the study at any time, and for any reason. That is, signing up for a study does not require that one complete the study if he or she becomes uncomfortable, or just changes their mind.

Another thing to keep in mind when people sign up for a study online is that the researcher has to assume that the individual is being honest in the information they provide. If for example the researcher is only interested in females between the ages of 30 and 50, we have to assume that people that desire to be in the study are in fact females within this age group. Usually no verification is involved here.

YOU: *OK! I think I got it so far. Sample size is not just a number out of a hat, and finding suitable participants actually involves a lot of planning and strategy. This is really good to know, and so much different from how I used to think of stats!*

ME: So far so good! But there is one other topic related to sampling we should mention, and this is the type of sampling method the researcher will use in his or her study. There are actually several, and some have variations within them, but let's see what some of the more common ones are.

To begin with, many people consider random sampling to be the "best" method. The nice thing here is that everyone in the population has an equal chance of being selected for inclusion in the study. This greatly reduces any kind of researcher bias that might creep in to the work. For example, if I see a potential participant whose name is Jennifer, I might select her because my daughter's name is Jennifer! I know, sounds lame, but removing any kind of potential bias is a sound research practice.

You might also read in your text book about "convenience" sampling. This is where a researcher simply takes whoever is available to be in the study. One common example might be a man or woman at the mall who is stopping people to ask their opinion on different candidates for an upcoming election. Or, a woman at the cosmetics counter may want women to smell a new perfume and give their opinion on it. The main idea here is that anyone who is around and agrees is OK to include in the sample.

Still another form of sampling is purposive sampling. In the perfume example above, the survey taker may only be interested in middle age looking women, so she would not necessarily reach out to sample very young or very old women. The women she does select are chosen because they have a quality the researcher is interested in, and in this case it is age. For that reason, her selection is purposive.

Finally, in some studies where it may be difficult to find participants, a researcher may use "snowball" sampling. This gets its name from the analogy of a small snowball rolling down a hill, and getting larger and larger as it approaches the bottom. In a statistical context, this means that one person who is in the study may know of someone else that would also be a good candidate for being in the study, and contact him or her with the information. Again this is most commonly seen when the population is very small (say, only participants who have an IQ over 200 and play the oboe). I know, this sounds pretty restrictive, but it is up to the researcher to make the determination as to who should be in their study.

YOU: *Hey! I just remembered something that may relate to sampling. I read somewhere that outliers are data samples that are really different from the majority of the responses given by the participants. And while I am at it, I also heard that missing data values can cause problems as well. Am I on the right track here?*

ME: Excellent Question! Outliers are a part of data collection, and in virtually every study, missing values will result in incomplete information. Both of these issues that you wisely mention should be addressed initially when the research is being formulated, and again during the write up (see Chapter 4 for everything you will ever need to know about writing a statistical paper)!

Let's start with a definition of an outlier, which is not as easy as it may at first sound. Generally speaking, an outlier is a data value, given by a participant that is far away from the majority of other values given by other participants. For example, when measuring the heights of American men, values over 7 feet tall might be considered outliers. Notice that there ARE men at or above this value, but not many of them!

The problem arises when one includes outliers while computing the mean (arithmetic average) of a set of scores. The inclusion of the one outlier may drastically alter the mean of the distribution, even though it is still accurate. Here is an example: Say we have the following set of 5 scores for height:

5'6"; 5'9"; 5'3"; 5'5"; 5"9"

If we want to find the "average" height, we just add up these values and divide by 5, which gives us a mean value of about 5'6". But say we add one more score, a man whose height is 7'2". Adding this one score increases the mean by almost 4 inches! So we might reasonably ask if this unusual value should or should not be included in the data set. Notice that the calculation of the mean is mathematically correct and accurate. But does this really portray the average height in this small sample?

Your statistics book will talk about using Z scores to determine outliers (more on Z scores in the next chapter), but again the basic idea is just HOW FAR away from the average does a score have to be before it is considered an outlier?

Now, to your other excellent point on missing data! This is definitely a major issue for most research investigations. It can happen for one of two reasons: First, people start to do the study and provide the data but then drop out (called attrition), or people complete the study but for whatever reason do not answer all questions. Either way, the data contains records (scores from the participants) that are to one degree or another, incomplete. Given that this is an undesirable outcome, what choices does the researcher have?

Essentially our researcher can do one of three things. She can either completely delete any and all records (participants and their information) if ANY information is missing. The nice thing about this option is that ONLY COMPLETE records are available for analysis. The flip side is of course a (sometimes significant) loss of participants, and a much smaller sample size.

The second option is to keep all of the records, but instruct the statistical software to only use complete data for the various analyses. Here is an example of how this might work: Let's say that Joe given us his age (as we requested) but not his income in dollars for the last year (also requested, but Joe did not disclose this). Now, if we want to find the mean (average) age of the participants, we can use Joe's information since we do know his age. But, if we wanted to find the average income of the participants, we would have to omit Joe from the calculation (and reduce the sample size) because there is no value to include.

The third option is to ESTIMATE the missing value, insert that estimated number, and run the calculations. Say that without Joe's income, the average income turned out to be $54,000 per year. We can substitute this value (called imputing the mean) into Joe's blank area for income, along with any other people who also did not supply this information to the researcher. Other estimation procedures exist, and perhaps you will run across some of them during your study of statistics.

YOU: *I had no idea that there was so much more to statistics than just formulas and numbers! I think I also heard you mention earlier about data that is valid and reliable? Can you explain what these really mean, and how they are different?*

ME: Absolutely! Reliability and validity are measureable features of a data set, but we can start with conceptual definitions of them. Reliable data are those that are reproducible over time. For example, I can get on my take my 9-iron and hit a golf ball 100 yards. I hit another ball with the same club and it goes 96 yards. I hit another and it travels 101 yards. As you can see, repeated "measurements," in this case yardage, tend to give reliable results.

But are the results valid? Ah, now that's another question. They may or may not be. Continuing with our golf example, say I measured the distance I hit the ball using a tape measure. My golf pro comes up and tells me that this is not a very good way to accurately measure the distance! She takes out a pocket digital yard measuring device, and asks me to hit three more shots with my 9 iron. She reads off the distance as 87 yards, 70 yards, and 81 yards. These much more accurate measurements of my distance are VALID measurements, as they are giving readings that are in line with the true reality.

Notice that something can be reliable yet not valid! Say my pros gadget is defective, and is adding 9 extra yards to my shots. I hit three balls and she reads my distances as 90, 90, 90. This is perfectly reliable (since all the scores are the same), and at the same time completely invalid (as her measuring instrument is off on each reading by 9 yards)! The TRUE distance is 81 yards, not 90 yards, so the device, while reliable, is not valid.

For instruments such as IQ tests, the SAT or GRE, personality inventories and so on, valid and reliable data are required if we are to make accurate assessments and offer useful advice from the results of the data analysis. Have you, for instance, ever taken an exam in a room that was noisy? That was too hot or too cold? That was stuffy or smelly? If you did, do you feel your scores are an honest reflection of your knowledge or abilities? In other words, are the scores valid?

Validity and reliability can be calculated (usually using some form of correlation), and as you might expect there are several different kinds of validity and reliability! Your statistics book my talk about split half reliability, or Chronbach's alpha as a validity coefficient, but remember in general what these mean (the conceptual definitions) and the number you see will have much more meaning.

YOU: *I was always wondering about the differences between those two concepts. Thank you! Any additional words on issues in statistics?*

ME: Actually, yes! I mentioned earlier about the idea of both *knowing* the specific assumptions for a specific statistical procedure, as well as *knowing how to test if* the assumption is met. Of course, we should also be aware of some things we can do if assumptions are *not* met! Let's see what is happening here.

Your text book and your instructor will probably emphasize many of the most commonly used statistical procedures. There are of course variations on these statistical themes, and sometimes an investigator or researcher must use very esoteric, specialized methods to analyze the data. But the main point here is that we cannot just enter a bunch of numbers into our software or our calculator, and churn out meaningful statistical results. The old saying "garbage in, garbage out" is very appropriate here! And remember that your software and your calculator do not know if the numbers you are feeding it are meaningful, valid, and appropriate! Only you, with your human brain can answer that!

To understand and appreciate the first assumption we need to refresh our memory on scales of measurement. As a short preview, Nominal Scales are just categories (types of cars), Ordinal Scales are rankings (first, second third), Interval Scales have equal intervals between values, and Ratio Scales have a true zero point (again, see below for more details).

Our first assumption is that the data are scaled on at least an interval level (meaning interval or ratio scaled data) is necessary since we are doing arithmetic operations on the numbers, and only numbers that meet this requirement can be validly assessed and interpreted (but re-read the section outlining why some experts feel that Likert scales are often OK even though they are not interval level).

Again, shortly we will be discussing the thrilling topic of scales of measurement (see the next section below), where we will learn that interval scales have equal spacing between values, but lack a true zero point. For now, the best example of interval scaling is the ordinary room thermometer, where the 0 does not mean lack of temperature, but is an arbitrary anchoring of the scale. Compare that to a ratio scale, such as dollars in a bank account. Here the value of 0 literally does mean a complete lack of money, and is a real, not an arbitrary designation. Notice that many tests, including ones you yourself take, are assumed to be interval scaled (so, someone who scores a 0 on the exam does not have zero knowledge, and someone who scores 50 does not have half the knowledge of someone who scores 100. Interval scales cannot make comparisons like this; only ratio scales with a true and valid zero point can).

YOU: *OK, this makes sense. But how do I know what kind of scale the data are?*

ME: That is where your expertise and familiarity with the data comes in. If you are conducting your own research, you will determine ("and clearly write in your report," says Leslie) the nature of your data. Many times previous research will give you a hint as to the type of data used ("and that is one reason why a thorough literature review is required in many research projects, including a dissertation" Leslie chimes in!).

Assumption two is that the data form a normal distribution, or put another way, that the data are normally distributed in the population (in fact, you will probably see this drawing, and the related idea of Z scores, very often in your statistical lifetime!). What this means is that if we were able to take all members of the population and measure their level of the variable, it would form a normal curve. This is the familiar "bell-shaped" curve we see so often in statistics, where the average scores (values of the variable) are mostly in the center, and fewer and fewer cases are seen as one moves in either the positive or negative direction.

As a made up example, we might say that the average number of cereal boxes in a typical middle class home is 7. We would expect a lot of people to have 8 or 6 boxes, as these are close to the average. Fewer people would have 9 or 5, fewer still 10 or 4, even fewer still 11 or 3, and so on. The point is that as we move away from the mean the frequencies of the values seen become less and less. So what we end up with is a large "lump" in the middle of the distribution, and more of a trailing off as we move towards the ends of the distribution.

Your instructor may discuss one or more statistical tests to ascertain the degree of departure from normality (after all, no sample of data is going to be *perfectly* and *exactly* normal). You will then learn, basically, where we draw the line. That is, how far a departure from normal can we allow, and still consider the assumption to be valid?

A third common assumption in parametric statistics is that variables show a linear relationship to one another. You probably remember (fondly) those wonderful days in high school geometry, where your teacher pointed out the formula for a straight line, how to find the slope of the line, and where the line intersects the Y axis. Ah . . . the good old days, right?

But it turns out that teacher was on to something, at least as regards to common parametric statistical analysis. What is assumed is that Variable X will show a linear (straight line) relationship to Variable Y. This can be confirmed visually using a histogram (plotting points from X and Y and connecting them to see how approximately "straight" the line is). Now, if a plot of the variables does NOT show a linear relationship, we have a couple of options: 1) we can remove the variable from consideration in the study, 2) we can use a method that allows for nonlinear (sometimes called "curvilinear" relationship), or 3) we can transform the variable in some manner to "make" it linear! Your instructor and your text book will go into the details of how these procedures can be used, but for now understand that if the method you are planning on using assumes variables will show a linear relationship, it is best to graph them and make a reasonable judgment of whether or not the assumption is met.

A forth assumption is that the data come from a random sample of a larger, well-defined population. This is ideal for research that wishes to generalize the results of the sample to similar individuals (those in the larger population). However, we should be aware that it is not always possible to conduct a random sampling of the population. That in itself does not mean that research cannot continue, but it does imply that results are specific to the sample used, and should not be generalized.

A fifth assumption is that the responses of one participant are not in any way influenced by the responses of another participant. This is sometimes referred to as the independence assumption, and is related to the researcher having control over the individuals in his or her study. For example, if person A glances over at person B's answers and copies them, the responses are not independent. As another example, husbands and wives often share similar views that they have acquired from being together, and their responses may be seen as linked (yoked) and therefore not independent.

When one wishes to compare groups (as we do in t tests and in the analysis of variance, or ANOVA), an assumption of the equality of the variances of the different groups is made. While the *means* of the groups may differ, the *variances* are assumed to be the same. This particular assumption actually has a specific test that is done to check on the degree of congruence of the variances, and it is called the Levene Test.

To end this section, keep in mind that these assumptions are somewhat relaxed for non-parametric statistical analysis. Fortunately, most all parametric methods have a counterpart in non-parametric procedures, so if several assumptions are severely violated, the wise researcher looks for other forms of data analysis to use.

YOU: *Good deal! This section has really reinforced to me how important knowing and testing the assumptions of the various statistical methods is in order to generate valid and meaningful results. But didn't you have a little more to share on scales of measurement?*

ME: Yes! Thank you for asking. Scales of measurement are critically important as they determine what can and cannot be done with the variable in a statistical analysis. These scales, which you may recall can be classified as Nominal (categories such as gender, educational levels, or ethnicity), Ordinal (data that have order but the differences between data points are not equal). Nominal scales use numbers only as a designation to define objects. These numbers cannot be used to find means, standard deviations, or other such statistics. For example, we might want to tally the number of cars from different manufacturers in a used car lot. We might find there are 8 Fords, 4 Chevys, and 2 Toyotas (yes, it is a very small car lot)! It is true to say that there are 14 cars, but to say that the "average" car is 14/3, or 4.66 is obviously meaningless.

Now, ordinal scales also designate, but go further to give an order, or a ranking, of the data on a certain dimension. A good example is the order of athletes finishing in a swimming race. First, second and third have order, but the first place finish might be 2 seconds faster than the second, and the second might be 9 seconds faster than the third. What is lacking here is an equal spacing between the data points, something that is handled by interval scales.

So, next in line comes Interval data, which like ordinal has a definite sequencing, but in addition has intervals between different data points being equal. For example, the distance between an IQ score of 100 and 110 is the same distance as an IQ score of 133 to 143. The interval of 10 points is the same in both. While this is a nice improvement over the ordinal scale, the interval scale falls short in not having a true zero point.

Lastly, there is the Ratio scale of measurement, which is like the interval scale, but this time with a true and meaningful zero point. Common examples are physical indicators such as time, weight, height, distance, and so on. Notice that in the Interval scale the "zero" is at an arbitrary or convenient location, as opposed to a zero that indicates the complete absence of something. Having zero money means you have no money (ratio scale), but having zero degrees Fahrenheit outside does not mean there is no temperature! See the difference?

Now the reason this is important for you to know is that many kinds of instruments and questionnaires use a Likert type scale to record the data of the participants. I am sure you yourself have done this many times! Have you ever evaluated a hotel after staying in it? Or a meal at a restaurant? Or an attitude questionnaire asking your opinion on something? Chances are good that the scale was of the "On a scale of 1 to 7 where 1 means strongly disagree and 7 means strongly agree . . ." kind. This is an ordinal scale (can you state why), and most often referred to as a Likert, or Likert type scale.

YOU: *This is all well and good, but where are we going with it?*

ME: Here's the point. Strictly speaking, Likert scales and the data they produce do not meet the assumptions of parametric statistical tests. The reason is that they do not have equal spacing between the various choices (the increase in agreement between 2 to 3 is not necessarily the same degree of agreement between 4 to 5 or between 6 to 7).

However, when they are graphed with the frequencies of each of the choices (how many people selected 1, selected 2, and so on), they often form a normal shape, bell curve. It is this important curve, called the "Normal

Distribution" that is central to most of the parametric statistical methods you will likely encounter in your course. For that reason, since the distribution of Likert type scores is "approximately normal," they are considered by some to be appropriate for common methods such as the analysis of variance (ANOVA) and others.

YOU: *I am really learning a lot from this chapter and your amazing book! More! I need more! You said something about "data transformation." Is this like the Clark Kent transforming to Superman kind of thing?*

ME: Good one! And actually, yes! Just as Clark and Superman are one in the same (but look and act differently), so too are raw (non-transformed) data and transformed data! Let's see what this is and why it might be done.

Looking at the second question first, data is usually transformed when the distribution of raw scores do not show adherence to the shape of the normal distribution. Of course, we would first check for outliers and data entry errors (putting in 17 instead of 71 would make a big difference), but it may simply be that the data do not follow a normal distribution. In that case, we may decide to find another measuring instrument (and see if the data are more cooperative), or consider a data transformation.

YOU: *OK, so far so good. But what are the different kinds of transformations, and what do they even mean?*

ME: NOW you are thinking like a statistician! Great questions! There are (probably) an infinite number of mathematical transformations available. One extremely simple one is to add "5" to every score. Try this on a simple set of data; what happens to the mean? What happens to the standard deviation?

While the mean will change, the standard deviation will not (try it and see)! So, since the standard deviation reflects the *shape* of the data distribution, simply adding a constant to each score will not affect it. This is called a linear transformation of the data.

However, if we *square* each score, or take the *square root* of each score and recalculate the standard deviation, we will see that it has in fact changed! This is one example of a nonlinear transformation, which means that the shape of the original scores has also changed. Transforming each score by squaring it or taking the square root of it produces a new distribution that may well have a more "normal" looking shape. You would then conduct your statistical test, such as an ANOVA or t test, on the transformed data.

The other common transformation is the log transformation (short for logarithmic). This involves find either the base 10 log of each number, or the base e (natural log, or ln) of each number, and again once transformed, using regular parametric methods on the transformed values.

Your computer program for statistics will easily do this for you (probably unless the data set is VERY small you would not attempt this by hand), and afterwards will save the new variable (the transformed scores are indeed a new variable) for later analysis. There are also a number of web based sites that can do the transformations if you insert the original scores.

YOU: *I see that it can be easy to transform the data set, and that the reason one might try to do this would be to alter the shape of the original distribution to one that has features of a normal distribution. But how are the results interpreted? What would the "Mean Log of something" actually mean?*

ME: Yes, this is a problem! Here is what is actually done. First, the decision is made to transform the data is made (and the reason, or reasons), for this are clearly explained and documented. Then, the data is transformed and saved, and finally the parametric analysis is done. As usual, a decision is then made based on a critical value, and the null is rejected if the value of the statistic fall into the rejection region.

THEN, the data is "back-transformed" by doing the inverse mathematical operation on the statistics (such as the means and standard deviations) and report that information in the original units. This is not "cheating," as you have stated what you are doing, why you are doing it, and offering an interpretation based on back-transformation to the original units. Keep in mind that this is a somewhat advanced topic, and may not be covered by some books, some instructors, and some courses.

YOU: *Great! Hey, I need a break! Any final words before we conclude the chapter?*

ME: Yeah, we did get somewhat wordy here, but let's end with one last concepts you need to appreciate, and that is model building.

YOU: *I used to build models as a kid! Race cars, planes, stuff like that.*

ME: Me too, but here we are talking about statistical (mathematical) models of the world as we see it applied to our research questions. For example, let's say we want to do a simple analysis of variance to check for mean differences in the dependent variable. To make this more concrete, say the dependent variable is "number of putts sunk on the putting green in 30 minutes," and the independent variable is the kind of putter a golfer has (the brand).

So, we might have 30 golfers with 3 different kinds of putters, and want to see if the average (mean) number of putts sunk from a fixed distance (say, 15 feet) differ among the three groups (10 in each one, each with the same brand of putter). We give each golfer twenty tries of sinking a putt from a fixed location, and record how many times each person makes the putt. After the last person has finished, we would have 30 scores (number of putts sunk) and could find the mean number for each of the three groups separately. The one-way analysis of variance (ANOVA) would then tell us if there was or was not a significant mean difference between the groups.

If you think about this a bit, we see that we have an implied model: Number of putts made successfully is a function of the type of putter used.

This in fact might be stated as a null hypothesis: Ho: “There will be no differences in the mean number of putts sunk by groups using different brands of putters.”

Now it may occur to you that this model is too simplistic. For example, better golfers may well make more putts regardless of the brand of putter they use. So, you decide to improve your model by suggesting something like “Ability of the golfer, AND brand of putter used will determine the number of putts made.” Here, you might see, we have TWO independent variables, (ability of the golfer, and brand of the putter), and with this model we can formulate hypotheses for a two-way analysis of variance.

The point to take away is that research questions and hypotheses imply an underlying model of the reality one is trying to understand. Keep this in mind as you are challenged with text book and class room exercises, and perhaps one day, your OWN research data!

Chapter Six

The Chapter of Announcements

OK . . . for those of you who have not yet had the good fortune to have me as your instructor in a statistics class, one of the things my students love most about me (as you might guess . . . there are several), is that in my online classes, I write a LOT of announcements and send out a LOT of emails. My purpose in doing this is often to clarify important concepts, emphasize aspects of the readings, or point out common errors. While I do use these methods in my traditional face to face classes, I find they are much more important in the virtual environment, where student and professor usually never meet.

Below you will find some of my best announcements from my classes in basic, advanced and multivariate statistics (remember that YOUR instructor may have their own specific goals, outcomes and procedures). There may also be more references and examples to psychological aspects of statistics, rather than business or biology, but really the comments and ideas can be generalized to all people in a variety of different majors, expect that parts of these will be familiar and make sense, but other parts may be fuzzy or vague. Don't sweat it if you don't get it (hey, I just made that up and it sounds good! Feel free to borrow it)! You will as time goes on. So dig in and see what you think.

MY TOP 8 RECOMMENDATIONS FOR DOING WELL IN A STAT CLASS

Hi everyone, and again welcome to the class. I am glad all of you are here with us. Here are some comments and recommendations for you to be as successful as you can in the class we are about to undertake together (and of course, you ARE reading my announcements, right?):

1. Plan to spend sufficient time on the readings, homework and other assignments. Most will probably find that about 10 to 20 hours a week is reasonable. This class will definitely require a major time commitment on your part, so be ready! I have found in the past that the most successful students, whether in an online or a traditional class, are the ones that give a high priority to time commitment.
2. Many statistics text books are quite challenging. Those who take the time to read the material slowly, carefully, and sometimes repeatedly REALLY benefit and understand most of the material. Those who skim it, or do not start early in the week, will find it frustrating and often complain it is a "bad" book for the class.
3. If offered, use the services of a Supplemental Instructor, a TA (teaching assistant) or a tutor for your class. Past experience from students in previous classes showed that they felt that the time invested here was VERY well spent.
4. Start early in the week to read the assignments. This one is REALLY important for success in the class. Find out what chapters are required by looking over the syllabus, and find the homework to get familiar with it. Be sure to click on and use the links we provide, as there is a lot of useful information here. Read the assignments carefully to be sure you are using the right variables and procedures.
5. Read all of my announcements. There is a lot of very helpful information, advice and instruction given to you. I also send occasional class emails, so check your email box for new messages from me. Use email to me if you have a request, concern, or a private issue. Please use the email from the class room, so that all messages remain here in the class.
6. Check into the class at least every other day. New announcements come often. I may change some small parts of the class, such as a discussion question change or a homework modification. It is your responsibility to read my notifications of these changes. If you do the wrong work, I will ask you to redo it.
7. Write your posts, response posts, and homework papers in a clear, organized and professional manner. Do not write as a casual conversation, write as a doctoral student in a respected Ph.D program. I do count the overall quality of your writing as part of your grade on all submitted work. Please do not use the "this learner" or "this researcher" format . . . speak in first person for postings.
8. For those taking statistics as a part of their psychology degree, get a copy of the new APA 6th edition manual, and read it. Some chapters

deal specifically with statistical output, presentation, and graphical displays.

Thanks everyone! See you in class.
—Dr. Trunk

A FEW INTRODUCTORY COMMENTS - JUST FOR YOU

Hello class. As we have learned, sets of scores (or distributions of scores, same thing) are simply numbers that have been measured and recorded on a set of participants. These measurements are assumed to reflect qualities of the individual, given that the test or inventory used was valid, and that the participantss (if they are human) answered honestly. While often we are dealing with questionnaires or written assessments as our measuring instrument, notice also that any equipment (such as medical equipment) designed to measure a feature about an individual must also be valid and reliable (if the calibration is off, all measures will be off as well).

But what do we DO with these sets of numbers? How are they to be analyzed, interpreted, and useful to us? Without some strategy for presenting the data and investigating what it might imply, all we have is a table full of numbers!

Statistics is actually a branch of mathematics, designed specifically to answer these kinds of questions. Forgetting for a moment about all of the other considerations of a professional research design (sampling, testing, qualitative or quantitative, and so on), statistics allows us to 1) See what we have, and 2) Interpret what we have. The former is usually called Descriptive Statistics, and the latter Inferential Statistics.

Descriptive statistics summarize information for us. If you were looking at heart rates for 50 patients, these 50 numbers would be very hard to make sense of; imagine if you had 500 or 5000 numbers in your data set! One thing we can do is produce graphs, which show us in a visual way what kinds of patters might exist. For instance, are most of the scores in the middle? On the high end? The low end? Are the scores clustered around the center, or spread out in the ends (tails) of the distribution? If we are looking at PAIRS of scores (such as heart rate and blood pressure), we can graph them on a 2 dimensional plane and see how one variable seems to relate to the second one. Do high scores on heart rate tend to go with high scores on blood pressure?

In addition to the graphs we can produce, we can calculate measures of central tendency (average) and measure of dispersion (variability) for our data. We know that the mean, median and mode are all useful for getting a feel for the average score, and that the range, standard deviation and variance tell us about the amount of variability in the scores as a whole. Note that these are simply summaries of the data; we are not concluding anything past the data, or making inferences about other individuals not measured in the original data set.

However, we are usually NOT just interested in the original set of data. We want to know if high heart rate, IN GENERAL seems to go with high blood pressure, or if low IQ IN GENERAL seems to go with poverty, or if training on a new method IN GENERAL seems related to improved performance on some other measure. These are questions of inferential statistics, and there are dozens of methods of analyses to answer these and related questions.

Many inferential methods look at group comparisons. In its most basic form, if we take one large group of people, randomly split them in two smaller groups, give one of these groups some kind of treatment (independent variable) and leave the other one alone (called a control group), and then measure them later, we expect to find differences between them. Since the only thing (theoretically) that was different was the treatment given to one and not to the other, we can infer a casual conclusion. Statistical analysis allow us to see if this conclusion was based more on the treatment, or on just dumb luck (chance). This is where hypothesis testing comes into the picture, along with type 1 and 2 errors. More on this (somewhat confusing) topic tomorrow! Thanks everyone.

Our goal in this introductory course is to familiarize you with basic research design and statistical analysis. While these two aspects of a research study are intricately intertwined, we often look at them separately. Here at Capella University, as well as at many other colleges and universities around the world, students in psychology take one or two courses in research design (methods), and two or three courses in statistics. Since our time is short for this class, I want to at least mention some of the more commonly seen statistical techniques, ones you may well encounter in your own research (or use in your own thesis or dissertation).

1. t tests are used to compare two different groups. t tests are actually the most simple form of analysis of variance (ANOVA), the difference being ANOVA looks at more than two groups. Independent t tests look at different groups, while paired t tests look at the same individuals at two points in time.
2. The one way ANOVA looks at one independent variable (such as religious preference) and one dependent variable (such as donations to charity). Independent variables are always qualitative (nominal, categorical) and dependent variables are always quantitative. The two way ANOVA adds a second independent variable, such as gender, or level

of education, or ethnicity. In these studies, we can also look at the interaction effects, if any, between the two independent variables.

3. Regression can be simple linear (one predictor) or multiple (many predictors) on some kind of quantitative dependent variable. For instance, age, number of previous heart attacks, smoking, blood pressure and number of medications might be able to predict the number of years a person can be expected to live. These analyses tell us which predictors seem to be important, and how much they are able to predict the outcome measure. There are many variations of this method, which we study in more advanced courses.
4. Factor analysis is a method of reducing a large amount of data into a smaller amount, without a substantial loss of information. It is looking for underlying commonalties among variables. These are sometimes called factors, or dimensions. For instance, intelligence is a dimension, based upon scores from several kinds of tests (these are the variables). How many different kinds of intelligence are there? What variables "load" highly on what factors?
5. Discriminant analysis and Logistic Regression both try to predict a dichotomy. For instance, the same variables that we see in the regression example (3) can be used to predict if one probably will have a stroke, or will not have a stroke. Notice here we are looking at a dichotomy, whereas in multiple regression we are looking at some kind of a numerical outcome.
6. Non parametric methods are used if the data do not meet the assumptions of parametric statistics (such as all of the above). They are sometimes called distribution free tests.

There are of course many variations on the above, but I hope this gives you a bit more of an idea about the multitude of ways that one can conceptualize a study, and analyze the data. Thanks everyone.

—Dr. Trunk

VARIABLES . . . OUR FRIENDS IN RESEARCH

Class, there is often some confusion on the distinction between quantitative and qualitative variables, so I want to address it right away. This is a central designation between variables, and determines the kinds of analyses that are permissible. Note that this is NOT the same thing as a quantitative or qualitative study!

There are TWO kinds of variables, qualitative (also called nominal or categorical) and quantitative (also called continuous or numerical). They must be analyzed TWO different ways.

1. Quantitative variables have scores, numbers or other meaningful measures that are numerical. Examples are age, income, GPA, final exam grades, and SAT scores. Notice that these variables are considered to be at the interval (or in some cases) ratio level. In most SPSS files, you will see several different kinds of these variables. To find descriptive statistics for them (means, standard deviations, ranges, medians, modes, and so on) use Analyze, Descriptive Statistics, Descriptives. Move the variable(s) you wish to get statistics for over into the main box, and then hit Options. Select the descriptives you wish to see, then hit continue and OK to see the results. This will produce a table with the requested statistics.

You can do a similar analysis by choosing the Frequencies or Explore procedures. Click on statistics and select the ones you want see. Look at the tables here and those for descriptives to see the similarities and differences. The output tables may look different, but the calculated values will be the same. Always check the Options tab, as SPSS has some nifty choices for the curious student!

Remember that the main idea is to summarize a list of numbers that have meaning, such as income, IQ, quiz scores, ages, and so on. Calculating descriptive statistics such as means and standard deviations is the starting point for most forms of inferential analysis. Procedures such as t tests, ANOVA, regression and correlation use many of these preliminary calculations, and expand upon them (take a look at the various formulas for some insights here)!

Now, Qualitative (categorical) variables can also analyzed by the frequency procedure. DO NOT INCLUDE SUMMARY STATISTICS, AS THESE ARE MEANINGLESS. Do not select ANY of the descriptive statistics under the Statistics tab; just run the procedure. Recall that qualitative variables are "dummy coded" with 1's and 2's, or some such system, purely to IDENTIFY the kind of variable (1 = males, 2 = females; 1 = Hispanic, 2 = Black, 3 = Asian, 4 = White, 5 = Other ethnicity). This lets SPSS count the number of each gender or ethnicity, and to find percentages relative to the total sample size, but NOT to find the "mean" of gender or the "average" ethnicity!

Let me elaborate a bit. Say in a particular data set we code males as 1, and females as 2. This allows SPSS to count how many males and females are in the sample of data. Take a look:

1, 1, 2, 2, 1, 2, 1, 2, 2, 2, 1

When SPSS see this, it knows that there are 5 men (since there are five "1's," and 6 women. But if you ask for the mean of these numbers, SPSS gives you a meaningless answer of 17/11, or 1.55! Obviously this has no interpretation, and is not a valid statistic, BUT SPSS WILL DO IT IF YOU ASK IT TO! Therefore it is up to YOU to be sure you know what you are doing!So, for any analysis you complete, do NOT find statistics such as these for qualitative, categorical variables. Instead, you will find frequencies and percentages (what percent of the total sample is female? How many males are in the sample?). These kinds of questions are meaningful, and you will generate tables that answer them.

Thanks everyone.

—Dr. Trunk

IMPORTANT COMMENTS ON HOMEWORK

Class, In addition to your discussion posts and responses, and your quizzes, each week you will have an assignment (I may refer to this as homework, or your homework assignment) that is to be turned in by Sunday midnight (your local time) of the assigned week. Assignments must be written in WORD, and must be placed into the correct drop box area. Here are a few reminders about the work:

1. Some of these assignments require you to enter the data into SPSS first, then do the required exercises. Not all of the homework requires hand entry into SPSS, but usually it does. For instance, for THIS WEEK, questions 9 and 10 on the homework require entry of the blood pressure data into SPSS first, then you can find the answers to the questions.
2. For each weekly assignment, some questions are optional, meaning you can try them if you wish, but do not have to. I treat them as extra credit . . . they cannot hurt your grade, but they can help your grade. So by all means if you have some idea of the answer, please include it.
3. PLEASE ANSWER THE HOMEWORK QUESTIONS IN THE SAME ORDER THEY ARE ASKED. It is OK to first copy the question, and then answer it. Please do NOT change the order of the questions, or I may miss one or more of your answers, get hopelessly mixed up, and consider moving in with Dr. Kostere to learn more about qualitative analysis.
4. Give me a sufficient answer to demonstrate you understand the question. Don't be overly wordy, but don't be so brief that I really don't know if you know or understand the answer. Example: If one has more than two means (two groups), what parametric test is probably indicated? Student A says:"The analysis of variance (ANOVA) is a parametric test that looks at more than two means to see if there are statistically significant differences between them. With a single independent variable, the one way ANOVA is best. Post hoc tests can be used if omnibus significance is found." Student F says: ANOVA. Would YOU grade these the same way?

5. Usually I want to see SPSS output copied and pasted into the WORD document. Have the tables near your discussion of them. DO NOT HAVE 14 PAGES OF TABLES FOR ME, THEN START TO DISCUSS THEM ON PAGE 15. Consider the reader, and do not make him or her scroll all over the place. Discuss and interpret the table NEAR the table, not 13 pages away.
6. Have at least a one paragraph introduction to the assignment. This should be the first thing I see. For example, for the Week 1 work, you might say something like "The first assignment involves several exercises that look for an understanding of basic statistical concepts. What follows are my answers to these questions." DO NOT START YOUR WORK WITH PAGE AFTER PAGE OFTABLES, CHARTS OR OTHER SPSS OUTPUT. Just as you would find this confusing, so will I, so please always provide an introduction to the reader to acclimate him or her as to what they will be reading.
7. You do not have to include any Venn diagrams. You may do so if you wish, but I am not going to require them and you will not lose points for not having them in an assignment.
8. Usually, I will grade and comment on your homework by Wednesday of the following week. If possible, I will get it back earlier to you, and if there are very busy weeks for me, I may be a day or two later than Wednesday.
9. I will send out a MODEL ANSWER each week for you to compare your work to. I may send out a "Mystery Student" answer as well, so you can see how others in the class approached the homework for the week.
10. START EARLY. You will fall behind if you don't begin the readings by Monday or Tuesday, and the assignment by Wednesday. Remember, in advanced graduate statistics, and for that matter in any statistics class, the bar is high, but definitely within reach.

Thanks everyone.
—Dr. Trunk

A FEW WORDS ON SCALES OF MEASUREMENT

Hi Everyone. Hey, I know when I think of Thanksgiving, I think of scales of measurement . . . so let me pass a few thoughts onto you.

It is very important to remember that scales of measurement must be appropriate for the statistical tests that you or another researcher are doing. In fact, many software programs, including SPSS (Statistical Package for the Social Sciences) will perform operations on the variables you tell it to, and generate impressive looking results. The problem is that the results are total garbage. Unless the data produces numbers that can be used appropriately and meaningfully in the analysis, the output is nothing but junk.

Here is an example that many new students often get confused with. Say you are looking at differences in reaction time (the dependent variable) between men and women (gender, the independent variable). Reaction time is a RATIO scale since it has a true, non-arbitrary zero point. It does not matter that no one can have a zero reaction time, it is still anchored at zero seconds.

Now, SPSS must know how to tell the difference between the score of a woman and the score of a man. So we use a code to tell the program this. For example, we can use 1 to designate that the responder is female, and 2 to designate the responder as a male. If we had three women and three men, SPSS would see this as

1,1,1,2,2,2.

So the first score is a woman, the second is a woman, the third is a woman, the forth is a man, and so on.

Say that the reaction times for these six (N = 6) people were:

23, 21, 25, 30, 32, 24.

This means that person one (a female) took 23 seconds to respond; person 5 (a man) took 32 seconds to respond, and so on.

Now, we can find the mean (arithmetic average) for women by adding up the scores and dividing by 3: 69/3 = 23.

The mean for males is 30 + 32 + 24 / 3 = 28.67 (notice that it is fine to have decimals here). We could (and will) see if the statistically significant differences in the average scores of men differ from the average scores of women by using the independent samples t test (but you will just have to wait a few weeks for this . . . unless you want to read ahead)!

These numbers "make sense" in that the means came from real scores (reaction times). But we CANNOT do the same for our Nominal independent variable (gender). It makes NO SENSE to say that the mean of gender is 1 + 1 + 1 + 2 + 2 + 2 = 9/6 = 1.5. The "number" 1.5 is meaningless, since gender is a nominal variable.

Say you had 5 telephone numbers . . . could you add them up and divide by 5 to get the "mean" phone number? Of course not! But SPSS will still do this analysis if you ask it to!

So be very aware in your readings to separate out Nominal (categorical) and Ordinal (ranks) variables from Interval (equal distances between scores) and Ratio (equal distances plus a true zero point). Only interval and ratio data can be analyzed by parametric statistical analyses (but see the discussion in the previous chapter about the use of Likert type scales in parametric methods).

Thanks everyone.

—Dr. Trunk

WHAT? YOU WANT TO KNOW ABOUT TYPE 1 AND TYPE 2 ERRORS? READ ON, STATISTICAL WANDERER (OR WONDERER!)

Class,

Type 1 and Type 2 errors are often a source of confusion for the beginning statistics student. Both are highly important, and as you will see, changes in one type of error result in changes in the other type of error. Let's see what they are all about:

Type 1 Errors:

When we postulate a null hypothesis, we are stating that there is no relationship, or put another way, no difference between means. Now, in reality, this is either a true statement, or a false statement. The problem, of course, is that we do not know which of the two is true!

So, we conduct some kind of a test. In the simplest case, we might look at two groups, and compare their average (mean) scores. Again, our null is that there is no difference in the means (or put another way, the mean of the first group minus the mean of the second group is zero). It is very unlikely, however, that the actual difference is exactly zero. But how far away from zero is it? If it is not very far, then (probably) our null hypothesis is true; and if it is very far from zero, then (probably) our null hypothesis is false.

Well, as it turns out, there is a cut off point for "how far from zero" the two scores have to be before we are reasonably sure that ARE different from each other! This is usually called the "critical value" of the statistical test, and depends on the sample size and the actual test.

Notice that there are two ways we can be "right" in our decision. If the means are really not different, and we conclude that they are not really different, we have made a good choice. And, if the two means really ARE different, and we conclude that they in fact are different, we have also made a good decision. This second kind of good decision is actually called "Power," and is simply the probability of rejecting the null hypothesis when the null is actually false.

Now here is the tricky part. Let's say that the actual value of the test is in the critical region (that is, it is telling us that the difference is far from zero). We need some kind of decision rule, and a reasonable one is that if the value is in the critical area, we will "reject" our null hypothesis of no difference, and conclude that the two are in fact different by a greater than chance occurrence.

But are we SURE, absolutely SURE, that we have made the right decision? In fact, we could be wrong! So we give this chance of being wrong a probability, we call it alpha, and we set the value to be .05. This means that 5 times out of 100, we will conclude that there IS a difference in the means, when in fact there is NOT a difference. This is called a type 1 error, and it is a pretty standard feature of most statistical tests involving differences in means.

Type 2 Errors:

So, a type 1 error occurs when we decide that there is a difference between means, but in reality there is not (just a chance occurrence). But, there is another kind of error we can make. Let's take a look at this one.

If we run our test and find that the two means are different (that is, the difference is far from zero and lies in the critical area), our decision rule says to reject the null in favor of the alternative hypothesis. But again, we could be wrong in this decision. Remember that just because we have decision rule we are infallible in making decisions!

So, a type 2 error (sometimes called beta) is the probability of rejecting the null when in fact the null is true. Fun, right?Z tests (and t tests) must have a dependent variable that is interval or ratio level. We cannot really do these kinds of tests with nominal or ordinal data.

ALL ABOARD FOR Z SCORES

Z scores, also called standard scores, are a way of finding how close to, or how far away from the mean, (average of all the scores) a particular individual score happens to be. A Z score of 0 means that the score is equal to the mean. If the Z score is greater than 0, the score is greater than the mean, and of course if it is less than zero (negative), the score is below the mean. Now, the Z score indicates HOW MANY STANDARD DEVIATIONS ABOVE OR BELOW THE MEAN A PARTICULAR RAW SCORE LIES. So a Z score of 2 means that the raw score is 2 standard deviations above the mean, and a Z score of -0.86 means that the score is .86 standard deviations BELOW the mean (since there is a negative sign).

Notice that when we calculate Z scores, we can compare two (or more) different scores (that is, from different distributions) to see which is "better" (higher relative to the other). This is because Z scores are directly comparable, while raw score distributions are not.In fact, we usually translate raw scores into what is called the standard normal distribution, with a mean of 0 and a standard deviation of 1. Using this common scale allows us to compare different test inventories. It does not matter if we are talking about dollars from housing sales, calories from Turkey dinners, or grades on a statistics test. Once all scores are converted to standard (Z) scores, they are directly comparable to one another.For example, say Joe took a test in History, that has a mean of 68 and a standard deviation of 2.4; Neal took a test in Psychology that has a mean of 89 and a standard deviation of 5.1.Who did better: Joe, who got a score of 72 on the History test, or Neal, who got a score of 95 on the Psychology test?

Who can tell me the answer? (Hint: Find both of their Z scores, using the formula in the book . . . which one is higher?) HEY . . . does this sound a lot like one of your homework problems for this week?

SOME IMPORTANT WORDS ON CORRELATION AND REGRESSION

Class,

This past week we learned more about probability, and how the concepts of probability can be used in examining null hypotheses. If you have time, check out the following site for some useful (and fun) probability issues:

http://www.mathgoodies.com/lessons/vol6/intro_probability.html

We usually specify both null and alternative hypotheses in our work, but the null hypothesis is the one actually tested by different kinds of statistical analysis. We are (kind of) hoping that the null can be rejected, since our real interest is in the research hypothesis, or alternative hypothesis. We set alpha at .05 or .01, perform the statistical test, and then based on the results make a decision about accepting or rejecting the null. Of course, our decision can be wrong (Type 1 and Type 2 errors).

This week we change gears a bit, looking at correlational procedures. Correlations are widely used in psychology, and come in many different flavors. Here is a VERY brief list of what you should remember from this:

1. The Pearson correlation measures the linear relationship between two quantitative variables. Correlations are symbolized in psychology as "r," and always range between -1.0 and +1.0.

2. If the sign is positive, both variables tend to move in the same direction. Negative signs means they are moving in opposite directions. Correlations of zero do not have any sign or direction.

3. The strength of the correlation is its numerical value, regardless of the sign. Correlations of +.3 and -.3 are exactly equal in strength. The closer the correlation is to 1 or -1, the stronger it is.

4. Correlations imply prediction. Correlations of 1 or -1 indicate perfect prediction, while correlation of 0 indicate no prediction. Intermediate values means partial prediction of one scores on variable, given scores on the other.

5. If you square the correlation coefficient, you get a new statistic called the coefficient of determination. This value is the percentage of variability in one set of scores (one of the variables) that is a result of the linear relationship to the other variable. So, if r = .6, then .36, or 36% of the variability in the scores is accountable by the fact that the two variables share a relationship with one another.

Your SPSS assignment this week asks you to run a correlation between Gender and IQ. Technically this is not the best example, as gender is not a quantitative variable, but run the analysis as given. You can also run IQ and ADD scores, or ADD and GPA, or IQ and GPA and see if you can make sense of the results.

Correlation is closely related to regression, a topic we look at in more detail soon. Thanks everyone for your hard work! I am very pleased with the class (even those who misspell my name, the rare but fatal Type 3 error)!

—Dr. Trunk

WHAT? YOU WANT MORE ON BASIC CORRELATION? WELL . . . ALL RIGHT!

Hey . . . let's talk a bit more about correlation, OK? Correlation is a popular procedure that is used to find out about the linear relationship between two variables. Correlation is symbolized by the little letter "r," and is sometimes called "Bivariate" (meaning two variables), or Pearson correlation (after the founder of the technique). Here are some must knows about correlation:

1. It is a pure number . . . there are no units (dollars, feet, cc's) involved with it. It is calculated by the formula you can find in any statistics book, and on the web, and will ALWAYS be in the range of -1 to +1. There is no such thing as a correlation of 2, or -31, or 1.1. These do not fall in the range of -1 to +1. Notice however that zero IS a legitimate correlation.

2. It tells us about the STRENGTH of the linear relationship between two QUANTITATIVE variables. The CLOSER the value of r is to EITHER 1 or -1, the stronger the correlation is. Notice that two correlations who have values of -.4 and +.4 are EQUAL in strength, but DIFFERENT in direction (see below).

3. It tells us about the DIRECTION of two quantitative variables. POSITIVE correlations means that in general, the two variables are moving in the same direction, either both getting larger or both getting smaller at the same time. NEGATIVE correlations means just the opposite; as one gets larger, the other tends to get smaller. Negative correlations are sometimes called inverse relationships.

4. It tells us about the LINEAR aspects, not any kind of curve or other relationship. One way to think about it is that the larger the absolute value of the correlation (closer to 1 or -1) the more the points on the graph appear to be a straight line. In fact, correlations of 1 or -1 would graph as a perfectly straight line, and correlations of 0 would graph as a completely random scatter of points. When your instructor discuss the idea of "least squared regression," this is what they are alluding to.

5. Correlations are useful for prediction; looking at #4 above, we know that if two variables are perfectly correlated (1 or -1), then we have perfect prediction. Given one value of one variable, we EXACTLY know the value

of the other. On the other hand, correlations of 0 give NO PREDICTION at all; given one value, we know NOTHING about the other. Of course, correlations in between these limits have more and more predictive value. Remember however that correlations are never interpreted as percentages. For example, it is NOT true that a correlation of .6 means there is a 60% correlation!

6. Correlations NEVER mean one variable caused another to occur. Even perfect relationships do not imply causality. For instance, everyday that you wake up, the sun rises. These two variables are PERFECTLY correlated (every time you wake, the sun rises), but obviously have no casual connections. While studies involving correlation and regression MAY suggest a causal relationship, one needs to do an experimental study in order to validate this speculation.

7. If you SQUARE the correlation, you get something called the "Coefficient of Determination." This is a fancy way of saying that a percentage of variability in the scores of the second variable are understood (not error) as a function of the size of the correlation with the first variable. So, if two variables have a correlation of .5, then 25% (.5^.5) of the variance in one is due to its correlation to the other. Compare this to the information in point number 5 above.

8. Continuing the above reasoning, if 25% of the variability IS accounted for, then 75% IS NOT accounted for. This is called error variability. The 75% is called the "Coefficient of Alienation." Notice that the two MUST add up to 100% (variability accounted for plus variability not accounted for).

INFERENTIAL STATISTICS ANYONE?

Class,

One of the first things we should be clear on is EXPERIMENTAL versus NON-EXPERIMENTAL methods. The former are largely discussed in chapter 8, the latter in chapter 9. Think of experimental work as one where participants (subjects) are placed, or subjected to, some kind of condition. In the simplest experiment, we might have two groups. For instance, Group 1 gets a drug, and Group 2 does not. If people were randomly assigned to one or the other of these groups, and at the beginning of the experiment all had (say) high blood pressure, then at the outset the average (mean) BP for group 1 should be about the same as the average BP for group 2. So far so good.

If the drug is effective, then the BP of group 1 should be LOWER on average than the BP of group 2. If the drug is not effective, then when we take BP's again the two groups should be about the same, right?

In this kind of work, we use a statistical test called a "t" test (t is little case, not capital). The t test is a way to see "how different IS different" between the two groups. Using our example, say that BEFORE the drug the average of group

1 was 140 and the average of group 2 was 143 (remember, since people were randomly assigned to either group 1 (the treatment group) or group 2 (the control group) we would expect the two means to be about the same).

AFTER the drug, we find that group 1 now has a mean BP of 120, and group 2 has a mean of 141. Obviously, these are different, but are they different enough to be not just due to chance? THAT IS THE PRIMARY QUESTION OF THE T TEST. A t test ALWAYS compares two groups on something. We assume that the treatment will not be effective (something called the null hypothesis; more on that soon). Then we give the treatment (in this case the medication) and then see if the two groups are meaningfully different. If they are, we reject our null hypothesis (that the treatment is not effective) and conclude that it is indeed effective. But if the two groups are still similar after one receives the drug and the other does not, then we do NOT reject our null, and conclude the drug does not lower BP.

Just consider the logic. Start with two roughly equal groups, Do something to one and do nothing to the other. Then retest to see if the group means seem to be similar or different. IF THEY ARE DIFFERENT ENOUGH, WE HAVE A STATISTICALLY SIGNIFICANT RESULT.

OK . . . let's move on! Recall that t tests attempt to measure the differences between two groups. In our example, we actually had an independent groups t test, since people were assigned to either the treatment condition (receive a blood pressure medication) or a control condition (no medication). Since the groups were roughly equal in their average blood pressures at the start, any differences between them, if the drug is effective, should be due to the fact that they received different treatments (levels of the independent variable). Of course, minor differences between the groups will always exist just due to chance. The question really is, are the differences large enough so that it is probably not chance operating? Put another way, are the differences large enough so that it is probably due to the drug (the independent variable)? The t test allows us to answer this question.

Now, we frame things in social science research as hypotheses. We never, repeat never, prove anything, and you should not report results as having "proved" something. The reason we do not prove anything is that we are using probability, the language of science. Let me give you another example:

Say you are trying to determine if a coin is a fair coin. Fair coins have an equal chance (probability) of landing heads or tails on any toss. So let's assume the coin is fair, toss it several times, and see what might happen. Our NULL HYPOTHESIS is that the coin is fair; the ALTERNATIVE HYPOTHESIS is that the coin is biased. Notice that ONLY one of these can be true, and we need to determine which one is more likely to be true (notice I did not say certain to be true, just more likely). We decide to do this by flipping the coin 10 times and recording the number of heads and tails. If the null is true (we assume it is at first), then the number of heads should be about equal to the number of tails.

We flip the coin ten times, and get 6 heads and 4 tails . . . do we accept the null or reject it? Most would argue we should accept the null, since IF the null is true, THEN getting 6 heads and 4 tails is quite likely (no one expects EXACTLY 5 and 5, although that could happen).

But what if we got 7 heads and 3 tails? Is the null still one we should accept? What about 8 and 2? Or 9 and 1? Or 10 and 0? Does getting 10 and 0 prove it is not a fair coin? No, but is SUGGESTS THAT PROBABLY THE NULL IS NOT TRUE, AND PROBABLY THE ALTERNATIVE IS TRUE. This is not the same thing as "proving" the coin is not fair; after all, it CAN happen that a fair coin gives 10 heads and 0 tails. It is UNLIKELY, but POSSIBLE.

We need a criteria to decide whether or not to accept or reject the null; we use the term ALPHA to give us the degree of acceptable error in our decisions. Almost always, you will find alpha to be either .05 (meaning that 5% of the time we will reject the null, when in fact the null was true); or .01 (which means that 1% of the time we will reject the null when the null is true). This type of mistake, rejecting the null when it is true, is called a TYPE 1 ERROR.

So why not just make alpha equal to 0? The reason is that MAKING ALPHA SMALLER MAKES ANOTHER TYPE OF ERROR MORE LIKELY (in statistics, just like in life, there is no free lunch). This other error is called BETA, OR A TYPE 2 ERROR. This is the error of NOT rejecting the null when the null is FALSE.

This is always confusing, so let me restate it: TYPE ONE ERRORS MEAN THAT THE NULL IS TRUE, AND YOU HAVE REJECTED IT. TYPE TWO ERRORS MEANS THAT THE NULL IS FALSE, AND YOU HAVE NOT REJECTED IT.

So like life itself, there is no free lunch! As the probability of one error goes down, the probability of the other error goes up. Remember that alpha is always preset; we usually do not worry about Beta, as it is a function of other parameters. On the other hand, the statistical power IS quite important, and is equal to (1—Beta). Power is a percentage of how likely one will conclude there IS a difference between two means, when in reality there really is a difference! Power is good in the statistical sense.

Just as we got a result for our coins, and made a decision based on how "rare" the outcome was, we do the same thing for a t test. We look at the differences between means, and make a decision based on how likely or unlikely those differences would be if the null was true. In our blood pressure example, there will be a point where the difference in the average pressures for the two groups will lead to rejection of the null (and by implication, the drug is effective in lowering BP). The details of the t test do not concern us in this class; but the results and interpretation do!

If the value of t is such that it would occur 5% of the time or less, we decide o reject the null. Notice we can be wrong here (a Type 1 Error). If the value of t is such that it is greater than .05, we decide not to reject the null; notice we can also be wrong here (a Type 2 error).

Now, that wasn't so hard was it?

YOU WANT IT? YOU GOT IT! STATISTICAL SIGNIFICANCE, THAT'S FOR ME

One thing I find often, and many students find somewhat surprising, is that the CONCEPTS of inferential statistics are as hard to grasp as some of the formulas and calculations involved. While statistics is a mathematical subject, specialized computer software, such as the Statistical Package for Social Science (SPSS) has taken the drudgery out of doing complex calculations by hand. Still, we need to know where those magical results came from, so of course some contact with numbers is required.

But numbers have meaning, and without being able to properly interpret their meanings, all the fancy calculations, graphs and tables are worthless marks on paper. SPSS and other programs put out a LOT of numbers, only some of which we really need to concern ourselves with.

One of the most important issues we deal with in statistics is whether or not our hypothesis about the world seems to be supported by the evidence, or not supported. We never have proved anything, and we are really not looking for proof. We are looking at probability, and if the probability we find indicates one decision or another is likely to be true. We are taking a snapshot of the world, which is just an indication of the truth, not proof of it.

Now you may not realize it, but YOU use probability all the time, every day, probably 95% of the day! You just don't think about it (it is unconscious), but it is there. Say you decide to cross the street; you are really calculating the odds (probability) of being successful (not getting hit, not getting caught by a cop for jaywalking); you taste your soup and decide that a bit more salt is needed - this means that if you add salt it will probably taste better. You dress up nice for a date or an interview for a job; you will increase your chances (the probability) the desired outcome will occur.

In most social science experimental studies we use the null hypothesis to indicate a non-significant result, and the alternative hypothesis to indicate something IS significant (which means that it is probably not due to chance. Our research question reflects the expectation of the alternative hypothesis (after all, if we did not expect to find a conclusion, we would not do the study in the first place)!

—Dr. Trunk

WELCOME TO THE CAFÉ STATISTIC—MAY I TAKE YOUR ORDER?

HEY! I'm hungry! Want to join me for a bite? I know a great place, just around the corner.

Hello, and welcome to the Café Statistic. My name is Barry, and I will be your server for this evening. May I start you out with a plate of t tests, or perhaps a nice correlation?

No? Well, then what can I do for you?

Ah . . . an ANOVA. Excellent choice. And as you may know, we have several varieties for you to choose from.

For the more straightforward in your party, I suggest the one way ANOVA. This popular choice comes with one independent variable (but several levels of that variable), and one dependent variable, which of course is interval level and quantitative. Of course, you may order a side dish of one or more or our tasty post hoc tests if your meal is significant.

Now for those with a slightly richer appetite, I suggest the two way, or Factorial ANOVA. Here you get a double portion of independent variable, and the usual one dependent variable and side order of post hoc tests. But for those interested in interaction effects (and who isn't these days, if I may be so bold), this is the ANOVA for you. One of our most requested, it goes well with a nice glass of wine (White Zin is always a good choice).

We also offer the three way ANOVA, but quite honestly not many people order this; seems that those pesky three way interactions are just too hard to digest. But of course, you are welcome to try it if you wish.

What? YES! Of course we offer covariates. How nice of you to inquire. We have a scrumptious analysis of covariance (ANCOVA, also makes a great personalized license plate!) with your choice of one or more covariates, which you specify as quantitative measures that will be removed statistically from your dependent variable. Following this removal (our chefs are EXPERTS at this!) we will proceed with your ANOVA; I am sure you will find the result to be to your liking, as your dependent measure is now free of possible correlational effects from the covariates. Please allow additional time for this order.

Feeling REALLY adventurous this evening? May I suggest our SPECIALITY, the multiple analysis of variance (MANOVA), or for those who just INSIST on covariates, our piece de resistance*, the MANCOVA. This full course statistical delight has all the features of our regular ANOVA, but two or more dependent variables may be specified. This is exactly what many sophisticated customers of the Café Statistic would like, but we suggest that you fully sample our many regular ANOVA's before you wrap your arms around this one!

Please relax, take your time, and I will be back soon. Thank you.

—Barry

*Pièce de résistance is a French term (circa1839), translated into English literally as "piece of resistance", referring to the best part or feature of something (as in ameal), a showpiece, or highlight. It can be thought of as the portion of a creation which defies (i.e. "resists") orthodox or common conventions and practices, thereby making the whole of the creation unique and special. The phrase gives the sense that the referred-to element is the defining essence of the whole, that part that makes it memorable or gives it its unique character.

PRACTICAL AND STATISTICAL SIGNIFICANCE—WHAT'S THE DIFF, DOC?

Class, most of our work has been focused on the p values, which indicate rejection of the null (statistical significance) or not (non-significance). However, just because a test of the null has obtained STATISTICAL significance does not mean it has any PRACTICAL significance.

As an example, say we measure 10,000 men and 10,000 women on a reaction time test. We find the mean reaction time for men is 4.31 seconds, and for women 4.30 seconds.

With a sample size like this, a t test would report a statistically significant result, even though the means differ by .01 seconds! We have to be very careful not to overestimate the practical significance of a study, even if it has statistical significance.

Conversely, many studies that do NOT have statistical significance have a great deal of practical importance. For example, knowing that two different drugs do not differ in their effectiveness in treating something is VERY important to know! One obvious conclusion is that if it makes no difference which one is used, perhaps the less expensive one to manufacture would be preferred.

Thanks guys! Hang in there for me for our final two weeks!

—Dr. Trunk

ANCOVA AND THE TIGER

People in advanced statistics often have problems with the Analysis of Covariance (ANCOVA) design. Really it is a nifty way to remove unwanted variance from a quantitative dependent variable. In fact, if you have ever gone bowling and used your handicap, or done the same in golf, you are (sort of) used the ANCOVA logic. Let's take a look!

I know . . . I know . . . Tiger has had "issues" recently, but since he is still one of the greatest golfers in the world . . .

Imagine a group of golfers:

Fred (the rat man) Skinner, who averages 85 for 18 holes.

Jean (the thinker) Piaget, who shoots 94 on the average.

Albert (the Imitator) Bandura, the best in the group at 78 strokes.

As they are getting ready to tee off for a relaxing round of golf, an athletic, handsome, trim, well dressed golfer runs up to join the group. His name is Tiger (the wishful thinker) Trunk, and he shoots 68 on the average for this course.

As they are getting acquainted on the first tee, Fred brings up the issue of a little wager on the round. Being gentlemen who have been known occasionally to bet a dollar or two on a round of golf, they decide to make a bet. Lowest score for 18 holes wins 5 dollars from the other three players!

Tiger Trunk, the greatest golfer in the world, is very happy about this wager! He is just about guaranteed to win! How can the others even compete?

Skinner, Piaget and Bandura, having learned of their incredible playing partner, are not too thrilled at playing the great Tiger Trunk straight up, as they would most certainly lose!

"The contingencies of reinforcement are just not there for me," says Fred.

"I cannot assimilate an even match into my cognitive schemata," moans Piaget.

"Yeah; me too! Like they said," echoes Bandura.

Being adept at statistics as well as golf, the incredible Tiger Trunk suggests an adjustment to everyone's final score, based on how many strokes difference there are in the averages. So, Skinner would get 17 strokes (85-68=17); Piaget gets 26 strokes, and Bandura gets 10 strokes off his total score. Everyone agrees that this is much more fair, and off they go to the first tee.

At the end of the round, Skinner shoots 84; Piaget ends his round with a disappointing 101; Bandura is really on his game (imitating the swing of the phenomenal Tiger Trunk) and winds up with an impressive 72! Tiger Trunk shoots his average score of 68.

Now, who is the winner? Why? What does this say about ANCOVA (adjustment of scores on the Dependent Variable to account for linear relation to another IV)?

What does this say about the fantasy life of your instructor?

—Dr. Barry (Tiger) Trunk

PLANNED COMPARISONS? POST HOC ANALYSIS? CAN YOU HELP ME HERE, O WISE PROFESSOR?

As we are learning, ANOVA is an omnibus (overall) test of the collection of means. It we have a one way ANOVA, we have one IV, but it may have several levels. Each level has a mean, and the null states that there are no differences in the means.The two way ANOVA, a logical extension of the one way ANOVA, has two IV's, and there is a null for the first one, a null for the second one, and a null for the interaction effect. But still the results of the ANOVA basically state that somewhere the means either are (or are not) significantly different from one another.

Now, if we do have a significant difference, we can do ADDITIONAL STATISTICAL TESTS (yeah, man!) to find out WHERE the differences lie. These can be specified either BEFORE the ANOVA is done, or AFTER it has been done. The former are called planned comparisons (since we plan them out before the ANOVA), and the latter post hoc comparisons (post hoc meaning "after the fact.").

As an example using the one way ANOVA, pretend we were comparing the average GPA's of four groups (freshmen, sophomores, juniors and seniors) and got a significant result. This must means that SOMEWHERE these four means show a (probably, since we can make a type 1 error) non chance difference; at least one pair of means is significantly different.

But WHERE is the difference? SPSS has two ways to answer this:

1. SPSS can do planned comparisons (also called contrasts, or sometimes planned contrasts), which are set up BEFORE the ANOVA is run. For instance, we might be interested in how Freshmen and Sophomores TOGETHER compare with Juniors and Seniors TOGETHER. Or, we might want to compare Freshmen to Seniors.

2. Post hoc tests can be used AFTER a significant result has been found. There are many different ones to choose from (see Field for a good general discussion), but usually we see the LSD, the Tukey, the Sheffe or the Bonferroni as the ones chosen. Usually one does EITHER planned comparisons OR post hoc tests, not usually both!

STATISTICAL SIGNIFICANCE?
A LITTLE MORE CLARIFICATION, PLEASE PROF

Now how does all relate to the concept of statistical significance? Basically, what is means is that an observed effect (result of a statistical test) is so large (different from what we expected if the null is true) that there is a very small probability we got it by chance or by accident (that is, we do not think that the result we got was due to chance alone) is called statistical significance.

How do we decide if something is statistically significant? We decide this by "drawing the line" between what we consider to be probable and improbable. Then we see which side of the line we results fall on.

If Ho (shorthand for the null hypothesis) is true, the p-value (labeled "sig" in SPSS, and is the actual, calculated probability value) is the probability that the observed outcome (or a value more extreme than what we observe) would happen. The p-value is a value we obtain after calculating a test statistic. The smaller the p-value, the stronger the evidence against the Ho. If we set alpha (the probability of a type I error) at .05, then the p-value must be SMALLER than this to be considered statistically significant; if we set alpha at .01, then it must be smaller than .01 to be considered statistically significant.

Remember, the p-value tells us the probability y we would expect our result (or one more extreme) GIVEN the null is true. If our p-value is less than alpha, we REJECT THE NULL and say there appears to be a difference between groups/a relationship between variables, etc.

Conventional results for statistical significance are usually given as $p < .05$ and $p < .01$; but what exactly do these mean?

$p < .05$ can be interpreted as "if we were to do this experiment an infinite number of times, this result would happen no more than 5% of the time (so 1 time in 20 samples), if the null were true." The same interpretation is give for a $p < .01$, except that of course the probability is 1/100.

What if your p-value is close to alpha, but slightly over it (like .056)? You cannot reject the null. Some authors will state a result like this as "marginally" significant, but APA does not subscribe to this kind of designation. You should not call results marginally significant in your write up s.

In addition to tests of significance, you will often want to look at your effect size to determine the strength of the relationship. Often, a moderate to large effect will not be statistically significant if the sample size is low (low power). In this case, it suggests further research with a larger sample.

Please remember that statistical significance does not equal importance. Something might be statistically very significant, but of no practical importance at all. You will always want to calculate a measure of effect size to determine the strength of the relationship. So, to reiterate, finding statistical significance is one thing; this does not imply any practical, or useful significance. That determination must be made by the reader of the article.

Chapter Seven

Odds and Ends

YOU: *I can't begin to tell you how my life has changed just from reading your book! Thank you Barry and Leslie for sharing your enormous knowledge with me!*

ME: Hey, no problem! In fact, since you have come this far, we want to mention a few other aspects that were implied in various parts of the book, and are worth some elaboration here. Is that OK with you?

YOU: *Absolutely! I'm all ears!*

ME: Let's take a closer look at, first of all, at statistical software. A bit of history: there was a time, not all that long ago, when students of statistics and mathematics were required to do problems by hand, using paper, pencil and a calculator (or a slide rule, if you go back a bit further)! While today we are not only comfortable with computers and software to take the drudgery out of certain tasks, the idea in years past was that learning was best achieved by doing the procedures, step by step.

In fact, let me share a secret with you: I agree! When I teach courses in statistics in a traditional format (as opposed to online), I often assign problems and ask for hand calculation. Given the formula, the data, and a basic knowledge of algebra, it is not unreasonable to ask college and university students to find a solution. And then, just for fun, we then enter the data into a statistical software program, run the analysis, and bask in the glow of success, as we see the answer WE arrived at is identical to what the software arrived at! Believe it or not, this is very fulfilling (in a geeky sort of way).

YOU: *So it would help me to better learn and understand a lot of the procedures by doing some of the problems by hand? Cool!*

ME: You got it! Start with some simple problems (like finding a mean or a standard deviation), and then move up to more challenging ones (like a correlation coefficient). Of course, there are some procedures that are simply too mathematically complex for most people to undertake without the help of a computer. Examples might be factor analysis, logistic regression, or canonical correlation. So let's not get carried away here guys! On the other hand, doing simple problems by hand IS a very good way of increasing your understanding of the work you are being assigned.

In this text I have mention one particular software program, the Statistical Package for Social Science (SPSS). The program, which a license must be either purchased or rented, is able to produce professional tables, graphs and other output that can be used in homework assignments, professional papers, or dissertations. Output can be formatted in a number of ways, and most all of the analyses are done using the point and click method with a mouse.

SPSS (and other software programs) require that the data either be entered manually, or imported from another source. Many very good books and websites exist to assist students with the details of entering data and running an analysis (but remember to check those assumptions first!).

So to sum it up, statistical software such as SPSS and other programs, is a tool to aid you in your statistical education. A very powerful collection of procedures that produce stunning, professional outputs, but more importantly, allow you to see the results of the analyses and make reasonable conclusions about the null hypotheses that translated the research question into one that can be answered statistically. Your instructor may require a particular program (Excel, SAS, and MiniTab are three popular ones), or you may find packages at retail stores or online that suit your purposes. But whatever product you end up using, keep in mind that *no amount of fancy statistical analysis or wonderful looking tables and graphs will substitute for sloppy work, for data that is inappropriate, or for careless data entry and screening.*

YOU: *Got it! What's next on the agenda?*

ME: Next on the agenda is the importance of data screening. This means not only taking care to enter the data correctly, but using procedures (and some common sense) to make sure that the data is valid for the analysis. To begin, keep in mind that often one must manually enter *thousands* of numbers into the statistical software. What would you guess are the odds that some are going to be entered incorrectly? For example, 17 and 71 are obviously very different, but SPSS will happily accept either one just as you type it in. As another example, .02 and 0.2 are very different values, but again SPSS has no what to "know" that a value is incorrect. As long as it is a number SPSS recognizes, it is good to go.

Clearly, for ANY data set, one must be very careful to enter the values appropriately. But there are additional ways one can check to see that this has been done. For small data sets, visual inspection is probably the first line of defense. Look for values that are impossible (like negative ages), and make corrections as needed. Decimal points and negative signs are often incorrectly entered, so careful attention is needed for data where these elements are present.

SPSS and most other programs can also check for values that are "out of range" in a set. For example, if you were looking at participants between the ages of 18 and 80, values below 18 or over 80 could be flagged and removed from the data. One way to do this might be to use an "if-then" approach, where the program searches for values outside of the allotted parameters, and upon finding them, highlights them for the researcher to review and revise.

While impossible values are a source of error, data values that are very much above or below the majority of entries (outliers) also need attention. While these values are not impossible or invalid, they DO tend to have a very strong influence on certain summary statistics, such as the mean. While identification of an outlier is the first step, the decision on what to do with it is perhaps more important. Of course, we do not just toss them out, or forget about them and just let them stay where they are! In fact, the researcher must justify his or her decision, and clearly communicate why and how extreme scores were included or excluded from the analysis.

YOU: *Excellent! What's next prof?*

ME: Next we take a quick look at pictures! You like pictures, don't you?

YOU: *Yes! Everyone does! In fact, my dad used to have this magazine, and ....*

ME: Uh, yeah. Why not let me tell you about statistical pictures, OK?

YOU: *Sure. Go for it!*

ME: "Pictures" are usually referred to as graphs or charts in statistics. These are visual ways of looking at data, and they can be very helpful! We can see things such as:

a. Trends over time
b. High and low points of the data
c. If the scores show a lot of variability (large variance) or are more compact,
d. Different locations for average scores (mean, median and mode)
e. If a linear relationship exists between two sets of data (correlation)
f. Percentages and frequencies expressed as slices of pie, as bars, or in other ways.
g. Easily find outliers, or other errors in the data set

The point is that graphs and charts allow the reader to see what the data look like when displayed in a reasonable manner. While tables are useful, they are often best used to summarize important statistics. Graphs on the other hand can show the entire range of data and reveal aspects that are not obvious or apparent in a table. For that reason, they are very common and fortunately, with today's modern statistical software, relatively easy to produce and customize. Your instructor may ask you to not only display your raw data in a table, but also in one or more graphs.

There are far too many tables for us to go over them in detail, but here are a few common ones with their proper names and most useful functions:

a. Histograms: These are graphs that are used with numerical data (that is, data that have real numeric meaning, as opposed to numbers that are simply used to categorize). The most common type is the simple frequency histogram, where bars that touch one another indicate the frequency of either an individual score, or a group of scores. This is seen by the height of the bars. The taller the bar, the more frequent the event. Variations on the simple histogram can also be used in certain cases for more clarity. One example might be a car lot where the owner wishes to display the various gas mileages of the cars in stock. Since gas mileage is a meaningful numeric measurement, a histogram is the appropriate graph.
b. Bar Charts: These are similar in appearance to histograms, but used for categorical data. For example, a car lot might want to show the frequencies of the various models of cars, such as Fords, Chevy's, Toyotas, Ferrari's, and Subaru's. The bars in a bar chart do not touch one another (as they do in a histogram), indicating that these are separate, non-continuous event. Like histograms, different versions and personalization options exist. Bars might be shown in color, or have patterns to make them more interesting.
c. Boxplots: Sometimes referred to as "Box and Whisker Plots," the graphs show the median value (the "middle" of the set of scores), with a box surrounding them. The top of the box represents the 75th percentile rank, and the bottom gives the value of the 25th percentile rank. In other words, the middle 50% of scores are contained in the box, with the median (the 50th percentile) right in the middle. Additionally, lines coming from the top and bottom of the box (called "whiskers") indicate the highest and lowest values! A great graph to show outliers in the data (which as we know have to be dealt with by the researcher)!
d. Scatterplots: This useful graph shows how various pairs of points are oriented in a plane (if there are two variables). For example, we might have data from 40 students on 1) the amount of time they study during the week, and 2) their grade point average. Probably these two variables are correlated (what direction do you think they would show? How strong a correlation might there be here?), and the scatterplot can help us to see what is going on. A modest, positive correlation (such as we might suspect here) would show a pattern of dots (pairs of data points) that slope from the lower left portion up towards the upper

right portion of the graph. Negative relationships would show the opposite pattern, and two variables with no linear relationship (a zero correlation) would just show a random pattern!

All graphs can be edited after they are produced. For example, one might make the font larger, smaller, or a different style. Color or shading can be added, and titles and legends (telling the reader what different symbols mean) can be inserted.

YOU: *Cool! I will be sure to add some graphs to my work to impress my professor with the incredible knowledge I have gained from reading this book!*

ME: A well-designed graph is an excellent addition in helping the reader better understand what you are trying to do and say. Are you up for one more topic and then we can call it a day?

YOU: *Sure thing! What is it?*

ME: Non-parametric statistics!

YOU: *Uh, OK. You talk, I'll listen...*

ME: Well, here's the main idea. When we work with most kinds of data, we are hoping that certain assumptions are met for a valid analysis. These assumptions (listed earlier for your reading enjoyment) include things like linearity, normality, equal variances, independent observations, and so on. But what if the data are not linear, not normal, or have very different variances (spread of scores) between different groups? What then?

When we have data that are not behaving nicely, we need alternative methods of analyzing it. These are called "non-parametric" methods, and offer ways to interpret data that do not conform to the assumptions of parametric methods. In fact, for most parametric procedures (like the t test, the ANOVA, or correlation) there is a counterpart in a non-parametric test! So, after checking the assumptions for the data you have collected, a prudent researcher makes an informed decision if the data should be analyzed in a parametric (assumptions met) manner or a non-parametric manner. So to summarize, one uses non-parametric methods when the data does not conform to parametric assumptions. Of course, no set of data will exactly conform perfectly, and most parametric methods are somewhat robust (resistant) to modest departures, but for data that severely violates the assumptions, non-parametric methods, like Superman, are here to save the day (or at least, save the analysis)! Let's take a quick look at some of the more common non-parametric methods, shall we?

YOU: *I can hardly wait...*

ME: I'll take that in the positive sense!

a. Chi Square: This test (pronounced "KAI-Square") has many uses. One of them is to test the relationship of two categorical variables (nominal scale) for independence. In more simple words, are two things related or not? For example, some people have dogs and some people have cats. Now within this group of pet owners, some people have Fords and some people have Chevy's. The question is whether or not there appears to be a pattern, or relationship, between the kind of pet one owns and the kind of car one drives. If there is, chi square will be significant and we will reject the null hypothesis of no relationship. On the other hand, a non-significant chi square implies that the kind of pet one owns is not related to the model of the car one drives. Note that there are many other uses for chi square, and in fact many statistical procedures produce a chi square (or pseudo-chi square) statistic! You instructor will happily dive into the details of these other uses for this non- parametric statistic.

b. Wilcoxon Rank-Sum Test and the Mann-Whitney test: Hey! Two for one! In fact, both of these non-parametric methods test the same kind of question. More specifically, do two different conditions, each with different participants, show similar results on an outcome measure? The main idea here however is that the outcome measure does not conform to the assumptions of a parametric method (like the independent samples t test), so a non-parametric method must be used. In these procedures, scores are rank ordered (lowest to highest), and we test to see if all (or most) of the lower ranks tend to belong to one group, while all (or most) of the higher ones tend to belong to the other group. If the groups really were NOT different, we would expect that some of the low and high ranks would fall into both groups. But if one group really did differ from the other, the ranks would "bunch up" to indicate this.
c. Wilcoxon Signed-Rank Test: This is the same as the above, except that the SAME participants are involved in both groups. Notice in the previous test that DIFFERENT participants were placed into the two different groups.
d. Kruskal-Wallis Test: When there are more than two groups, the Wilcoxon Rank-Sum test cannot be used. In this case, the Kurskal-Wallis test comes to the rescue, as it allows for several levels of one independent variable (so it is the same as the one way ANOVA, but used with the assumptions of the ANOVA cannot be adequately met).
e. Friedman's ANOVA: In this situation we again have more than two groups, and additionally we have the same participants in each group. So we might test people at time 1, time 2 and time 3 for differences in an outcome measure. Again, if the outcome measure does not meet parametric assumptions (as are required for the repeated measures ANOVA), we can use the Friedman ANOVA to see if differences are evident between the different time periods.

This is just a small sampling of non-parametric methods, but hopefully you get the main idea. They are all methods that attempt to answer the same research questions as their parametric counterparts, but do so using modified methods to account for non-normal, non-linear, and non-homogeneous variance data! I am sure you will sleep better at night knowing this!

Chapter Eight

Wrapping It All Up

Well friends, we have come to the last chapter of our little book. Has it helped you? Do you feel more confident, less anxious sand more ready to take that class and not just get through it, but really understand it? If so, we have accomplished our goal.

It really wasn't so bad, was it? In fact, once people get over their initial apprehensions about statistics, they often find it to be enjoyable (or at least tolerable), not only in their professional lives reading books, journal articles and dissertations, but in their everyday life as well. Statistics and statistical thinking is all around us, and perhaps now you will be in a better position to appreciate its contributions to your life and our world.

Statistics is much more than a bunch of calculations, tables, graphs and output. It is a quantitative, systematic method of answering questions that puzzle us. For most of us in the social sciences, we are more concerned with finding the answers to our research questions, rather than the details of the mathematics involved in the calculations of our statistics.

So while an understanding of the math is not our primary concern (but keep in mind in your particular statistics course, your instructor may have a different vision for his or her students, and may include more mathematical aspects), we should and must be able to decide which forms of data analysis best answer our questions (including qualitative methods!), if our variables meet the statistical assumptions of our procedures, if the null and alternative hypotheses have been stated properly and in a manner that their statistical answers can be interpreted in terms of the original research question, that tables, graphs, charts and other output are meaningful and correctly produced, and finally that we as researchers or consumers of research can critically think about, interpret and evaluate quantitative results.

But what if you are still somewhat uncertain? Still nervous and scared? Maybe reading the book again would help! But seriously, let's take another look at some of the things we have learned, and what Leslie and I suggest for you.

First of all, you should realize that when it comes down to it, statistics is just a tool. Statistics courses are designed to familiarize you with the tools of your trade, which in our case is professional research analysis in psychology. Do we want you to understand statistics? Yes, we do. Do we want you to become an expert in every kind of statistical analysis? Yes, we do (just kidding, we don't)!

But we do expect that you will 1) become familiar with the most common forms of analysis used in psychology, 2) be able to interpret most statistical results from professional journals, dissertations, and other scholarly works, 3) understand the basics of SPSS, and be able to use SPSS to enter actual data, and to run, analyze and interpret the results, and 4) know where to go for help. You probably have more sources than you realize: books, websites, colleagues, and alumni would be a good place to start.

The PhD in psychology is the highest degree conferred in our field. Graduates who have received this degree are expected by the psychological community to be well acquainted with research methodology, qualitative analysis, and of course quantitative analysis. While some PhD's do go on to specialize in the development and evaluation of new, complex statistical models, most do not. Most have the skills to adequately address points one through four in the above paragraph, and for the faculty at SOSBS, that is an excellent start to your professional life as a PhD psychologist.

Virtually all major universities stress critical thinking. What better way to demonstrate this ability than to dissect a research article. Instead of skimming over the results (or, heaven forbid, skipping them entirely), we can now say "OK . . . what are they trying to do? What are their variables? How were they measured? Were the measurements accurate, valid and reliable? Was the sampling technique adequate in size, and appropriate in design for the study? Does there appear to be good internal validity? What do the results indicate about the hypotheses of the study?" You get the idea. We are past the point of being spoon fed, and need to question, in a respectful and professional manner, the findings of a published article.

How many of you out there have found that learning statistics the second time around was easier for you? Why was this the case? Are the instructors at your school just so much better than ones you had in the past? Or perhaps you now have a level of maturity that allows you to easier and more effectively assimilate the information presented. Perhaps better statistical software, easier data entry, better text books and just more resolve on your part is the answer.

Don't be surprised if this does not continue for you well past your graduation from your particular school, college or University. Continue to read your books; look at chapters that were not assigned, "play around" with some simple data in SPSS, and see what happens when you do this or that option.

The world is complex, and as professionals we are trained not only in the content of our field, but in the methods of research design and data analysis. Our little book here is no substitute for a full fledged statistics text, but if we have inspired you, answered a few lingering questions, and given you the resolve to continue to learn and grow, our goals for the book have been met.

Just for the sake of closure, here is a brief recap of some important terms and concepts that you will be expected to know and understand when you complete your courses in statistics at most Colleges and Universities:

1. t test: A parametric procedure used to look at mean differences between two groups. Independent t tests use separate groups, and repeated samples t tests (also sometimes called dependent measures t test, or paired samples t tests) use the same people twice. Recall that parametric methods rely on data that are scaled on Interval or Ratio levels. The t test is not appropriate categorical data. One more interesting thing about the t test: the dependent t test can be use with samples that are assumed to be related, so that answers may not be independent of one another. For instance, husbands and wives, partners living together, and brothers and sisters all share a common environment, so scores on an assessment may be related due to this common feature. In that case, a researcher may decide to use the dependent t test, rather than the independent t test, if they do in fact feel the scores have a relationship to one another.

2. Null Hypothesis: The hypothesis of no differences between groups. Also, the hypothesis that a value (such as a correlation coefficient) is zero. This is the hypothesis that is tested with a statistical test. Actually, this is quite important, as students may feel that the statistical procedure is testing to see if the alternative hypothesis is true or not. Other kinds of null hypotheses exist, for example, the hypothesis that two distributions have equal variances. These are also tested, often when assessing potential violations of assumptions within a procedure, but most often a null refers to an actual statistical test of means.

3. Alternative hypothesis: The opposite of the null, what the researcher WANTS to find. This is usually a more formal statement of the research question. Keep in mind that the goal of statistics is to test the viability of the *null,* and to reject the null if the observed values fall into the rejection reason. This is NOT the same thing as *proving* the null is false, or *proving* the alternative is true. In fact, you will rarely, if ever, see the words prove or proved in a professional article. Researchers are humble! In addition, they recognize the results of any one study, no matter how elegant it may be, only SUGGESTS that a particular outcome is probably more true than not true.

4. Statistical test: The actual mathematical procedures involved in the calculation of a statistic. The observed result is usually compared to a critical value. If the result is beyond the critical value, the decision rule suggests that the researcher reject the null in favor of the alternative. If the result is not beyond the critical value, the decision rule would be to not reject the null. Another way of saying this is that if the result of the test is very different from what we expect, reject the null as being true in favor of the alternative (but see number 6 below)!

One common source of confusion that beginning statistics students have is knowing which of the dozens of tests available is the right one for them to use. Most statistics text books have some kind of "flow chart," or diagram that helps individuals to better determine the proper test for the research questions and hypotheses they are interested in. For example, the flow chart may ask "Is the data measured on at least an interval scale? (and aren't you glad you know what they are talking about)! If the answer is yes, the arrow leads in one direction (parametric methods) and if no in another direction (non-parametric methods).

5. df: Degrees of Freedom. How many observations are "free to vary" after the population mean (or really, any population parameter) is estimated. This value will differ for different tests, and is involved in finding the test results. The term is always written in lower case letters, and never as DF. The term "free to vary" can be confusing at first, so here is an example that might help: Say you are at a party and there are four people standing in front of four chairs. You tell the first one (Teresa) "Please sit in any one of these four chairs," and she does! She was free to choose and of the available chairs, and after having done so, you have three people and three chairs. Your next guest (Kim) takes one of the three chairs, and now there are two people and two chairs left. Your third guest (Susan) takes one of the two remaining chairs, and here is where "free to vary comes in: *After the one of the final two chairs was chosen by Susan, the final guest, George, MUST take the remaining chair.* He has no choice here, as there is only one chair left. So, the point is, out of the 4 original chairs and the four guests, the first three chairs (and guests) were free to choose, but the final one was not after the three had selected their chairs. The degrees of freedom in this example is three, since there were three choices that could be made (with the final choice determined, not free to vary).You will see this term often in your statistical education, but remember it basically tells you the number of independent choices available.

6. Type 1 Error: A mistake involving rejecting the null hypothesis, when in fact the null is true. This value is preset before an experiment, and given a probability called alpha (usually .05 or .01 by convention). The value of alpha is chosen BEFORE the statistical tests are run, and tell us the probability of incorrectly rejecting a null that should NOT have been rejected.

7. Type 2 Error: A mistake involving NOT rejecting the null when it is in fact false. A calculated probability called Beta can be done, but it is not preset like alpha. In general, if alpha is smaller, beta will be bigger. Your instructor and your text book will teach you the various ways to determine Beta, but essentially it is, like alpha, the chances (probability) of making an incorrect decision. Beta is the converse of statistical power.

8. Power: Power is a probability of correctly rejecting the null hypothesis when the null is in fact false. Power depends on the level of alpha, the effect size, and the sample size. All things being equal, larger samples result in greater statistical power. Power will also increase if alpha is made larger, and if the effect size (true differences between population means) can be shown to be large. Power is usually required to be at least .8 (meaning there is an 80% chance of finding a difference IF a difference really exists), but for many studies power may be as high as 95% to minimize a Type 2 error.

9. Correlation: A pure (unit free) numerical measure of the linear relationship between two quantitative variables. Symbolized by r, it ranges between - 1 and +1. Closer values to either of these limits indicates more prediction and more relationship, while correlations closer to zero indicate less linear relationship. Correlations provide both direction (positive or negative) and magnitude (strength) of the linear relationship between two variables. Keep in mind that some variables may show a NON-linear relationship, and that the familiar Pearson correlation would not be appropriate in this instance. In other words, the correlation may be close to zero because while there IS a relationship between the variables, it is not a linear relationship.

10. Coefficient of Determination: The square of the correlation coefficient. It tells the percentage of shared variance between two variables. Another way to look at this is that if two variables are redundant, r squared tells us this percentage. For example, if two variables have a correlation of .6, they have a 36% overlap. Notice that a correlation of .6 does NOT mean there is a 60% linear relationship!

11. Coefficient of Alienation: This statistics tells the percentage of independence, or non-shared variance, between the variables. This is also known as error. Note that this value is just equal to (1 – the coefficient of Determination)! Looking at the above example of a correlation of .6 and a shared variance of 36%, we see that the coefficient of alienation is just 1 - .36, or 64% unknown relationship. Not so hard, is it?

12. Analysis of Variance (ANOVA): This term applies to a variety of methods, all of which have one common purpose. Do real differences exist between the means of different groups? The groups are usually considered to be the Independent Variable, and there can be more than one. In that case, both the main effects of the independent variables separately, as well as their interaction, can be statistically evaluated. ANOVA can be One Way (meaning one independent variable, not just 'one way" to do the analysis), Two

Way (two independent variables), and Three Way (three independent variables, where this usually ends!). Keep in mind that two and three way have not only main effects that are tested, but various interactions (combinations of the independent variables).

13. Repeated Measures ANOVA: A form of ANOVA that looks at the same participants' scores at different points in time. Can be performed alone, or with one or more between group variables. The most simple form of this is the familiar "pretest – posttest" design where the same group of people take an assessment at two different times. Most often, some kind of intervention or treatment is given between Time 1 and Time 2, and the researcher looks for significant differences in the mean (average) scores between the two assessment times. In this analysis there can (and often are) more than just two time periods, and differences between any of them can be investigated.

14. Analysis of Covariance (ANCOVA): The main idea here is that we are doing an analysis of variance on a dependent variable that has been adjusted for the influence of a known covariate. A covariate is a quantitative variable that has a strong correlation to the dependent measure, but is not itself an independent variable and needs to be statistically controlled for.

15. Multiple Analysis of Variance (MANOVA): Analysis of correlated dependent variables among one or more categorical independent variables. Essentially the researcher here is interested in two or more related (correlated) dependent variables, and wants to see if there are group differences in their combined influence. Note that this is a different research question than asking if there are group differences in the two variables *separately*!

16. Multiple Analysis of Covariance: Identical to MANOVA, but scores are corrected for one or more covariates prior to the analysis. Recall that a covariate is a another quantitative variable that is thought to influence the outcome (dependent) variable, but is not one that the researcher wishes to study per se. Often the covariate is used as a correction, or adjustment, to account for differences in the initial group of participants.

17. Linear and Multiple Regression: A form of regression to predict a quantitative outcome measure from one or more predictor variables. The simplest format is one where there is one predictor variable and one outcome variable; this is called simple linear regression. If there is more than a single predictor variable (but still one outcome measure), we have multiple regression! Most commonly the predictor variables and the outcome variables are interval scaled, indicating meaningful numbers, but other variables (categorical) can also be used in regression if they are dummy coded.

18. Logistic Regression: A non-parametric form of regression that predicts a categorical (nominal) outcome measure from one or more predictor variables. The term "non-parametric" is sometimes used to refer to data that does not depend upon features of a distribution to be evaluated. This is sometimes called "distribution free" data. Most of the time the data involved

is nominal or ordinal (ranked). Compare to linear or multiple regression, where the outcome measure is numerical as opposed to categorical.

19. Discriminant Analysis: Similar to logistic regression, but follows parametric procedures. This method is more powerful than logisitic regression if the assumptions of parametric methods (such as linearity, normality, and independence of observations) are met. However, both methods often find the same thing and lead the researcher to similar conclusions and interpretations of the data.

20. Factor Analysis: A structure based analysis with many uses, including reducing a large number of correlated variables into a smaller set of factors without substantial loss of information. Factor analysis attempts to find order (structure) in a large correlation matrix by grouping similar items into a new entity, called a factor. A non-mathematical example might be someone who looks at their junk drawer (and let's face it, EVERYONE has a junk drawer!) and decides to organize it. Someone might put all the "school supplies" together, all the "electronic" stuff together, all the "kitchen supplies" together, and so on. In this case, these three Factors all contain related (correlated) items. Pencils and paper clips would be found in school supplies, and not in the electronic factor! Factor analysis can be either exploratory (looking for structure in a correlation matrix, and reducing the large number of variables to a much smaller number of factors), or confirmatory, where factor analytic methods are used to validate a theoretical structure of the variables.

21. Chi Square: A non-parametric analysis of the relationship between two (or more) categorical variables. Chi square can be used as a "goodness of fit" test to see how well a set of data approximates a known distribution, or can be used to find the degree of linear relationship between two nominal variables. For example, if we looked at a table that showed us the variables of political party (say Democrat and Republican for this example), and gender (using male and female for this example), chi square tests the null that there is no relationship between political party and gender. It does this by comparing the observed frequencies to the expected frequencies, and see how far apart these two are from each other. As if this weren't enough, Chi Square is also an important part of other statistical procedures, such as logistic regression. Many other non-parametric tests are available as well! In fact, some of you may take an *entire* course in non-parametric methods!

Well formerly anxious reader, you have made it to the end of the book! We hope that some of the ideas, comments and reflections in this book will assist you in your journey of becoming a professional researcher, or for obtaining your degree, or, for some, simply having less anxiety about taking and passing one or more courses in statistics. Remember not to be your own worst enemy when you do tackle a statistics course. Instead, keep an open mind, read everything your instructor provides, ask questions when you are confused or unsure of something, and give yourself plenty of uninterrupted quiet time to do your work. Of course, come back and visit us anytime! Re-reading section of this text can be very useful as a supplement to your required stat book.

Thanks everyone . . . see you online! And feel free to email us with comments, questions, or corrections!

Barry Trunk, Ph.D
Leslie Olsen, MA
November 1, 2015

Appendix A

A Complete Example of the Analysis of Variance

Let's take a look together at the Analysis of Variance, or ANOVA (always in Capital Letters). Our goals are as follows:

- Appreciate the application and usefulness of quantitative methods such as ANOVA, and the conclusions one can draw from using these methods.
- Determine whether ANOVA might be appropriate to answering your dissertation research question.
- Recognize several of the assumptions of the method, and some of the limitations of this approach.
- Better understand published research that utilizes the ANOVA approach.
- Review several of the variations on the ANOVA procedure, each applicable to a particular research question and design.

To assist with these goals, we will first present some of the theory and background of the approach, and then follow a completely worked through example of ANOVA. While moving through the example, you will be asked to calculate some simple results, and see how these intermediate results allow us to move forward to final conclusions about the hypotheses under consideration. Let's start with some of the basics:

- Analysis of Variance is one of many parametric techniques designed to look at differences between groups. Keep in mind that no matter what kind of ANOVA you might consider performing, the primary objective is to compare group means, and to see if these means are similar or statistically different from one another.

- If comparisons are made between different groups of participants, the Between Groups (Independent Groups) design is chosen. If the same participants are measured more than once, the design is called Within Groups (or Repeated Measures). It is also possible to combine these two into a Mixed Design. In this case, both between groups and within groups variables are analyzed.
- Most designs test a Null Hypothesis of no differences between means (averages) of the groups. The Alternative Hypothesis is that one or more significant differences between means are present between the groups.
- By convention, we use an alpha of .05 to test the null hypothesis of no difference. This means that we will incorrectly reject a true null hypothesis only 5% of the time (this is also the probability of making a Type 1 Error).
- ANOVA is an omnibus test; that is, it looks for differences in mean scores overall, but does not look (unless asked!) to find specific pairwise differences. If a significant result is found, one or more post hoc (after the fact) analyses can be performed to find out exactly which groups are differing from one another. Or, the researcher may perform planned comparisons, conceived before the data is analyzed. These planned comparisons look at specific pairs of means that are of theoretical or practical value to the researcher.
- Measures of association (eta squared) can also discover how strong the relationship is, given a significant result. Notice that a significant result may NOT be very strong.

Here are some additional features of the ANOVA procedure:

- The purpose of the Analysis of Variance is to discover if differences between groups exist. The means of each group will be compared to one another. These means are likely to vary from one another, hence the name analysis of variance (between means).
- Notice that it is average differences between Groups, and not between Individuals, that is the focus of the study. In these kinds of investigations, we are not usually concerned with individual scores; instead, we are comparing means of different groups.
- In a well-designed study (one with strong internal validity), the larger the differences between the means of the groups, the more likely that the independent variable(s) is exerting an influence. If the influence is large enough (relative to the influence of random error), we will be able to reject the null hypothesis of no differences between groups, and conclude that probably a difference exists. Notice we never prove a null or alternative hypothesis as true or false; instead we make a statement of probability to determine how likely or unlikely the hypothesis may be.

- The F statistic (named by Roland Fisher, and always a capital "F") is used to determine how large a difference exists between the groups. It is actually (as we will see) a ratio of two estimates of the variance in the scores of the dependent variable. If we are doing the calculations by hand, this computed statistic is compared to a "critical value" for a particular sampling distribution. If the computed F is larger than the critical F, we reject the null.
- If a software program (such as SPSS) performs the analysis, the computed F and the level of significance are given in the results table. In this case, we do not need to look up any values in a statistics book...the software does this for us!

Analysis of variance will "work" with any set of numbers. But data must be verified to insure that the assumptions of the procedure are met. Here is a brief outline of the most important assumptions of any ANOVA procedure:

- NORMALITY: Dependent variable scores in the population have a normal distribution, OR the sampling distribution of the mean is normally distributed. Using a histogram, one can take a visual inspection of the distribution of scores. More exact tests (such as the Shapiro) are available in SPSS, but visual inspection is always a good first start.
- HOMOGENEITY OF VARIANCE: The variability (variance) of scores under each level of the independent variable are equal. The Levene procedure tests the assumption of equal variances between levels of the independent variable.
- INDEPENDENCE OF OBSERVATIONS: A score (dependent variable datum) on one observation in no way affects any other scores on any other observations. All scores are independent of one another.

Note that the ANOVA procedure is somewhat robust, and that minor violations of these assumptions will not greatly affect the final results.

Many different kinds of ANOVA can be performed. The type of ANOVA that one uses depends upon the research question that is being contemplated.

The most basic comparison possible is between two independent groups, such as men and women, trained employees versus untrained employees, and so on. In this situation the analysis of variance is known as a t test. The t test looks at the differences in mean (average) performance of the two groups, and allows one to make a judgment if the differences are large enough to be statistically significant (probably not due to random chance). In addition to independent groups, t tests can also look at pre and post testing, where the same individuals are tested at different points in time, or a one sample t test, comparing a sample outcome to a well-known population parameter, such as an IQ score.

- A One Way ANOVA (of which we will see an example in the next few pages) looks at ONE independent variable and ONE dependent variable. The independent variable is categorical (sometimes called nominal), and may have several levels within it. The dependent variable is always quantitative, and measured on interval or ratio levels. For example, one might look at levels of stress (the dependent variable) in Freshmen, Sophomores, Junior and Senior high school students (the independent variable). Rank is the categorical variable, with four levels.
- A Two Way ANOVA has TWO independent variables and ONE dependent variable. It is possible that the two independent variables may interact with one another. Using the above example, we might add Gender to Rank, producing now eight separate combinations. Each main effect (Gender and Rank) will be tested, along with the Gender by Rank interaction.
- An Analysis of Covariance (ANCOVA) is one where the effects of a quantitative variable that is correlated with the dependent variable has been statistically removed before the ANOVA is performed. In our above example, we might feel that people with different GPA's might have different levels of stress. An analysis of covariance can statistically equate people on GPA (remove the correlation of GPA and anxiety), and then perform an analysis of variance on the adjusted scores.
- A Multiple Analysis of Variance (MANOVA) has one or more independent variables and TWO or more dependent variables. Continuing with our example, we might add measures of physical stress, and psychological stress as dependent variables. Multiple analysis of variance is an example of a multivariate analysis, covered in more detail in Section 3 of this Guidebook.

YOUR TURN: EXERCISE 1

Think of a research problem in your field of study where an ANOVA might be a useful approach.

Be sure to identify and clearly define:

1. The population you wish to study.
2. The type of ANOVA you might use.
3. The independent variable, including all relevant levels.
4. The dependent variable (remember, it must be a meaningful, quantitative measurement).
5. Any covariates that might be involved.

A Complete Example of a One Way ANOVA

Consider the following research problem:

The superintendent of a school district is trying to decide what method of teaching reading to first graders should be implemented in the 14 schools under her jurisdiction. She decides to conduct a study to compare three different methods of teaching reading using a One Way Analysis of Variance. Based upon the outcome of her study, she will have a more valid basis for her final selection.

Since she is interested in different methods of teaching reading, her independent variable is METHOD. Notice that the independent variable is categorical, and divides people into different groups, or levels. This variable has the following three (qualitative) levels:

1. The TRADITIONAL method of teacher/student instruction.
2. The VIDEO method of watching animated characters on a TV screen teach reading through songs, games and stories.
3. The COMPUTER method where children use simple software with color, animation and music to learn to read.

Of course, in a real study all of these variables would be clearly and explicitly defined, but for this example we will assume we know what they are. Since the dependent variable is quantitative, the superintendent will use a validated test of reading ability that will be given after 6 weeks of instruction. She decides to call this variable "Readscore."

She decides to randomly select 1 school from the 14 in her district. Then, she randomly selects 30 students from all of the first graders in this particular school. This procedure is called random sampling. (Note: In a real study, the number of participants would probably be much larger).

She decides to randomly divide the 30 children into three groups of 10 students each. This procedure is called random assignment to groups.

YOUR TURN: EXERCISE 2

Why does the superintendent need to use random sampling and random assignment to place the children into groups?

What are one or two advantages of using these techniques, rather than simply placing any child she wishes into whatever group she wants to place them in?

Does the superintendent have to get permission from the parents of the children, or since this is such a "gentle" study, is it OK to just use them without permission?

In keeping with standard, traditional methodology, she frames the following hypotheses:

NULL: There will be NO differences in the average reading scores among children in the three groups after 6 weeks of instruction.

ALTERNATE (Sometimes called the ALTERNATIVE): There WILL be differences in the average reading scores among children in the three groups after 6 weeks of instruction.

Notice that the NULL and ALTERNATE Hypotheses are Mutually Exclusive (only one is true) and Exhaustive (there are no other possibilities available).

While the Alternate represents the interest of the researcher, it is the Null that is statistically tested, and either accepted or rejected. Rejecting the null is not the same thing as accepting the alternative!

The raw scores from the 30 children selected to be in the study are shown in Table A.1.

YOUR TURN: EXERCISE 3

Calculate the mean ($\sum x / N$) for each of the three levels of the independent variable. What might you conclude from this quick comparison? Why is JUST finding the means of each group and visually comparing them not enough to render a decision?

The Sums of Squares:

In the ANOVA model, the sums of squares (individual score deviations from a mean, squared) are necessary steps in the final calculation of the F statistic. Your statistics instructor will show you the details of their calculations (bet you can hardly wait, right)?

Table A.1.

ID	TRADITIONAL	VIDEO	COMPUTER
1	15	22	12
2	10	19	18
3	17	25	13
4	20	14	9
5	16	20	11
6	20	23	10
7	11	18	14
8	19	24	15
9	12	18	10
10	10	17	8

For the one way ANOVA, there are three different sums of squares (SS):

- SS (Total)
- SS (Between Groups)
- SS (Within Groups)

It is true that SS (Total) = SS (Between) + SS (Within); these quantities will be displayed in Table A.2, the ANOVA Summary table.

Sums of squares are intermediate quantities necessary to calculate the F ratio.

Degrees of Freedom

Like Sums of Squares, the degrees of freedom (abbreviated as df, always in lower case letters) are necessary in the final calculation of the F statistic.

Think of degrees of freedom as the number of observations that are free to vary. For example, say we have 4 people and 4 empty chairs:

- The first person can sit in any of the four chairs.
- The second can freely choose any of the remaining 3 chairs.
- The third person can freely choose either of the 2 chairs remaining.
- But the LAST person MUST take the last chair! So, for these four people, there are THREE degrees of freedom.

In the One Way ANOVA, degrees of freedom are additive, just like Sums of Squares:

- df (Total) = N – 1 (Remember that N is the total number of participants.
- df (Between) = k – 1 (k refers to the number of groups, NOT the number of variables).
- df (Within) = N – kIt is true that df (Total) = df (Between) + df (Within)

YOUR TURN: EXERCISE 3

Given that N = 30 and k = 3, calculate:

1. df (Total)
2. df (Between)
3. df (Within)

Show that df (Total) = df (Between) + df (Within).

Combining the Sums of Squares and their respective degrees of freedom allows us to calculate a new statistic, called the Mean Square. By definition, the Mean Square (MS) is equal to the Sum of Squares divided by its proper degrees of freedom.

MS (Total) = SS (Total)/df (Total)
MS (Between) = SS (Between)/df (Between)
MS (Within) = SS (Within)/df (Within)

Note that while SS and df are additive, Mean Squares are not! In other words, MS (Total) ≠ MS (Between) + MS (Within).

The F Statistic: A Ratio of Mean Squares

We come at last to what we are looking for, the F Statistic (named after Ronald Fisher, the developer of the ANOVA).

- F is defined as MS (Between)/MS (Within)
- MS (Between) represents variability due to both group influences and random error; MS (Within) represents variability only due to random error.
- If the Null Hypothesis is true, there are NO group differences so the ratio of MS (Between) to MS (Within) should be close to 1.0.
- If there are group differences, then this ratio should be greater than 1.0. If the ratio is large enough, we will reject the Null Hypothesis of no group differences!
- F is calculated by SPSS and presented in a ANOVA Summary Table. Also contained in the Table are the various Sums of Squares, Degrees of Freedom, and Mean Squares.
- Note (The different ANOVA designs will have many different SS, df and MS; but the logic is exactly the same as in the One Way ANOVA).

The results of an analysis of variance are summarized in Table A.2.

YOUR TURN: EXERCISE 5

Verify that the Sums of Squares are additive, the degrees of freedom are additive, the Mean Square Between and Mean Square within are correct, and that the F ratio is the Mean Square between divided by the Mean Square within.

Table A.2. ANOVA Summary.

Readscore	Sum of Squares	df	Mean Square	F	Sig
Between Groups	326.667	2	163.333	13.047	.000
Within Groups	338.000	27	12.519		
Total	664.667	29			

Combining the Sums of Squares and their respective degrees of freedom allows us to calculate a new statistic, called the Mean Square.

- If you use SPSS to run the ANOVA procedure, the last column (Sig.) will tell you the probability of getting the presented results by chance, given that the null hypothesis is true. It is NOT necessary to look up values in a statistical table if SPSS or other software is performing the analysis.
- If the Sig. value is less than 0.05, we say that we have a statistically significant result. This means that probably, the results are not due to chance. Probably, the results are due to the influence of the independent variable on the participants.
- Another way to look at this is that Sig. is the calculated probability of a Type 1 Error, given that the null is true.
- In our example, a Sig. of .000 means that the obtained result has a probability of less than .001. It does NOT mean that there is a zero probability of a Type 1 error!

Post Hoc Tests

As we know, the ANOVA is an omnibus test, and looks for significant differences between ANY pairs of means. But if one wishes to look at each, separate pairwise comparison, one or more post hoc tests must be used. SPSS provides many different choices here, and Learners will have to read on what they do and do not offer. They basically differ in how conservative or liberal they are in their calculations of differences.

As mentioned earlier, a researcher may, at the outset of a study, have one or more planned comparisons of interest to him or her. These are decided upon before the results of the overall analysis are performed. Various methods of coding are needed for planned comparisons, and the interested Learner is referred to one or more of the advanced statistics books, given later in this guide, for assistance.

Summing it all up!

In this Appendix, we have learned that:

- The ANOVA approach is used primarily to see if differences exist between group means.
- Many different forms of ANOVA are available, depending upon the particular research question.
- The assumptions of the ANOVA should be met, but the method is robust to minor violations.
- The calculation of F depends upon a logical, linear sequence.

- The ANOVA Summary Table gives information on Sums of Squares, Degrees of Freedom, Mean Squares, F and the calculated level of Significance. Post hoc tests are available if a significant F value is found.

Now if you that the ANOVA was fun, buckle up as we move into Multiple Regression! Oh yeah!

Appendix B

Multiple Regression for You and Me

In Appendix A, we looked at the one way analysis of variance (ANOVA), and discovered how starting with raw data, were able to compute means, standard deviations, sums of squares, degrees of freedom, mean squares, and finally the F statistic. Comparison of the observed F to a critical value of F (based on degrees of freedom) will result in either the acceptance or rejection of the null hypothesis of no differences.

In this second Appendix, we turn our attention to Multiple Regression, a very general procedure that allows the researcher to make predictions on a quantitative dependent variable as a function of values on several predictor (independent) variables. Our goals are to become familiar with Multiple Regression, both in terms of general theory, and specific applications of the technique to research.

- Part 1: Theory and Fundamental Equations of Multiple Regression
- Part 2: Examples of Multiple Regression
- After reviewing the information here, you should be better able to:
- Explain the logic of multiple regression, and recognize how many research questions might be answered using these techniques.
- Understand that there are many ways to "do" multiple regression.
- Learn how SPSS can be used to run a multiple regression analysis.
- Explore the output and results of a multiple regression analysis.

We begin our discussion of this popular method of data analysis with the theory and fundamental equations of multiple regression. Multiple regression is a very general method that can be utilized to answer a wide variety of research concerns:

- Can a quantitative dependent variable be predicted based upon scores of several quantitative and /or categorical (dummy variable) independent variables?
- How much influence does a particular IV have on the dependent variable?
- How much additional predictive value is gained when additional IV's are added into the regression equation?

Understanding Multiple Regression is dependent on a solid understanding of several basic statistical terms and concepts. You should be familiar with the following:

- Measures of Central Tendency
- Measures of Variability
- Bivariate correlation (r)
- Equation of a straight line (slope, intercept)
- Hypothesis testing terminology (Null and Alternative Hypotheses, Type 1 and Type 2 errors, Power, Statistical Significance).
- Correlation & Regression are two sides of the same coin. If you study the actual formulas for both, you will see that several of the terms are identical!
- The degree to which we can make predictions is related to the extent to which variables are related.
- Correlation coefficient (r varies between -1.0 and +1.0).
- Regression Line (Line of "Best Fit" for the data).

YOUR TURN: EXERCISE 1

Discuss the relationship between points in the above scatterplots and the size of the correlation coefficient. What would a correlation of 1 or -1 look like?

Similar to the ANOVA procedure, Multiple Regression has several assumptions and considerations:

- Number of cases (participants in the study) (N): Approximately equal to 50 + 8k, where k is the number of variables. Too many cases or too few cases can lead to questionable results.
- Absence of Outliers: Remove, Rescore or Transform. Outliers must be considered in any statistical analysis.

- Absence of Singularity (correlations of 1.0 or -1.0) and Multicollinearity (very high correlations in the .9 or -.9 range): Makes the mathematics of regression impossible.
- Assumptions of Normality and Linearity of Residuals (differences between observed and predicted values): Tested through SPSS Examination of Residual Scatter plots.
- Variables should show a linear relationship to one another. This can be looked at visually by producing scatterplots of each pair of variables.

Just as if there are several methods of conducting an ANOVA, there are also several ways to conceive and run regression analysis. The researcher must carefully consider his or her design to best decide the kind of regression to run. Your statistics instructor, along with diagrams from your text book, will help you to better understand the differences. Basically, they involve how the variance (spread of scores) in the data are partitioned. Listed below are the three main forms of regression analysis:

In Standard, or Simultaneous Regression (also called Forced Entry), all IV's are entered at the same time, and each IV is evaluated in terms of the additional predictive power it adds independent of the others. Only the Unique, not shared contributions of the variables is considered.

A second type of regression analysis is called Hierarchical Regression. In this type of Regression analysis, the researcher determines the proper order of entry of IV's into the regression equation (as opposed to the forced entry method, where all variables are entered at the same time). Shared variance (correlation between the variables) is accounted for by the order of entry.

The third major form of regression analysis is called Stepwise Regression. In Stepwise Regression, the order of entry of the variables is determined by purely statistical criteria. The IV that correlates highest with the DV is entered first; of the remaining variables, whichever correlates highest with the DV (that is, add the greatest amount of additional predictiveness) is entered, and so on.

Notice that in this method, the software selects the variable with the highest zero order correlation to the dependent variable. In this case, Variable B has the highest correlation, so it is entered first, and areas 2, 3, and 4 are accounted for by the variable. The computer then compares the remaining variables for correlation to the dependent variable, and selects the larger of the two. In this case, it is Variable C. Finally, Variable A is entered last.

YOUR TURN: EXERCISE 2

In your own words, explain the differences between simultaneous, hierarchical and stepwise regression.

Multiple Regression is basically a method to generate a "best fitting line" to a set of data. The deviations from the line represent errors of prediction, so the goal is to minimize the deviations (which we square so that positives and negatives do not simply cancel to zero) of the points around the regression line. This is called Least Squares Regression, and is the most common form of regression analysis. It looks like this:

Yi' = A + B1X1 + B2X2 + B3 X3 + . . . BkXk

In this equation:

Yi' is the predicted value of the Dependent Variable for participant i.

A is a constant that orients the line in space.

B's are regression weights, which can either be raw score or standardized (Z score) weights.

X's are the scores on the (k) Independent Variables for participant i.

- This equation produces the smallest differences between the observed DV scores and the predicted DV scores, and for the entire sample is measured by the Multiple Correlation Coefficient, R.
- Once again, this is called "Least Squares" Regression.
- R, then, is the correlation between the observed and predicted dependent variable scores. In fact, if you use SPSS to run a Pearson correlation between the actual scores of the participants, and their predicted scores, the results will be R.

Regression weights can be raw scored or standard scored. If they are raw score weights, they represent the coefficients by which each participants original scores on the predictor variables (independent variables) are multiplied. They are generated according to the Least Squared criterion.

If all scores are first standardized (converted to Z scores), then the weights will also be in standard form, and will vary between -1 and +1. Since Z scores can be directly compared, so too can the size of the standard score regression weights. In this instance, the intercept (A) is zero, and the larger the coefficient of the variable, the more important it is in the equation. Again, Beta (b) weights will vary between +/-1.0, and are shown below:

Y' = b1X1 + b2X2 +b3X3 + . . . + bkXk

The "b's" are called beta coefficients, and are directly comparable to one another since they are on the same scale.

YOUR TURN: EXERCISE 3

Pretend that a researcher is studying the following variables related to gas mileage in new cars:

1. Weight of the car in pounds.
2. Length of the car in inches.
3. Cost of the car in dollars.
4. Power of the engine in horse power.

If these 4 variables are predictor (independent) variables, and average gas mileage is the criterion (dependent) variable, make a case for the researcher either generating a regression equation using raw scores, or Z scores. If you were the statistical consultant, what might you advise, and why?

So the takeaway here is to understand that multiple regression uses two or more predictor variables to estimate a quantitative score on some outcome measure. We have seen that there are different ways to "do" the regression (and we recall how exciting that was!). Next, we will take a brief journey into the wonderful world of multivariate statistics! Stay with me now, it won't hurt! Trust me . . . I'm a doctor . . .

Appendix C

Hey—Let's Look at Multivariate Statistics

In Appendix A, we visited the one way analysis of variance, a procedure that basically investigates mean differences between groups. While there may be several levels of one independent variable, or several independent variables, there was only one dependent variable that was the focus of study.

In Appendix B, the method of multiple regression was investigated. Here we saw how several predictor (independent) variables could be used to make the best quantitative prediction of one criterion (dependent) variable by forming a linear equation.

For both of these procedures, only one dependent measure was involved. Many different forms of data analysis do in fact have this basic structure, where only one measure is being looked at. These forms of analysis are collectively called Univariate Statistical analysis. But another large class of analyses can be found where either more than one dependent measure is investigated at the same time, or (sometimes) very advanced forms are considered. Experts have not reached a consensus on exactly what specific procedures constitutes Multivariate Statistics, but clearly they are ones that are built upon the foundation of more simple, univariate methods. In fact, we could say that common procedures like t tests, ANOVA or linear regression are special cases of multivariate statistics.

Several different multivariate methods can be used to help explain similarities of variables (factor analysis) or cases (cluster analysis), relationships between two sets of variables (canonical correlation), prediction of the odds of an event happening (logistic regression), finding weighted sets of predictors that can distinguish between two or more groups (discriminant analysis), and extensions of the t test and ANOVA to cases where two dependent variables are mathematically combined into one (Hotelling's T squared and multiple analysis of variance (MANOVA).

In this final Appendix, we will look at MANOVA, Discriminant Analysis and Factor Analysis. Our coverage of these very complex topics is meant more to acquaint you with their most basic features, to better understand how they are used in current research publications, and to encourage you to think about how they might be appropriate to your own research questions. Details of these procedures are explained in most college level statistics text books.

As we begin our coverage of multivariate statistics, keep these goals in mind:

- Appreciate the usefulness of advanced quantitative methods, such as Multiple Analysis of Variance, Discriminant Function Analysis, and Factor Analysis.
- Evaluate the advantages and disadvantages of Multivariate Methods.
- Determine when one or more of these techniques might be appropriate for answering your research questions.
- Become familiar with the assumptions and limitations of these approaches, and see how they are similar to the more familiar univariate methods.
- Be able to better synthesize and integrate published research that utilize multivariate statistics in their designs.
- Better understand the basic terms, concepts, assumptions and conclusions of several multivariate analyses.

The world is a complex place. You are a complex person. To fully understand you, we would have to know many different things about you at the same time. In the same way, multivariate statistics attempts to look at several different "parts" of a problem at the same time. Univariate methods may not adequately answer a particular research question, or may artificially investigate separate variables independently of one another, even if the variables themselves are correlated with each other.

On the other hand, to simply combine variables because one can do this is not a good idea either. Only careful thought and reflection on the research question, the variables and the rationale for treating them separately or simultaneously can result in a solid design. Here are some considerations that one should keep in mind when contemplating a multivariate analysis:

- For most analyses, additional time, participants and possibly money are needed. Inclusion of more variables means more instruments must be utilized and measurements gathered.

- A research question may be better addressed using a more basic technique; more complex methods of statistical analysis will not save a poorly designed study.
- The interpretation of the results can be challenging; e.g., if the variables of generosity, intelligence, musical ability and preference for isolation are all correlated, what does this mean? Finding a result and understanding a result are very different processes!
- For many analyses, a basic understanding of correlation and regression is necessary.

MULTIPLE ANALYSIS OF VARIANCE

Let's begin our brief foray into multivariate statistics with the multiple analysis of variance. The primary purpose of the Multivariate Analysis of Variance (MANOVA) is to investigate how two or more dependent variables are collectively related to group (independent variable) differences.

- In MANOVA, we are testing the null hypothesis that mean differences between groups on a linear combination of dependent variables probably did or did not occur by chance.
- Dependent variables are correlated with one another, and a new dependent variable that maximizes group differences is created.
- The new dependent variable is a linear combination of the original dependent variables, generated so as to separate groups as much as possible.
- Every participant receives a score on the newly created dependent variable (again, a linear combination of the original variables that takes into account their correlation, and configured to maximize group differences), and ANOVA is performed on this new dependent variable.

As in all ANOVA procedures, certain aspects must be adhered to. Some of these include:

- Participants are randomly selected from the population and randomly assigned to groups.
- Multivariate Normality exists (The sampling distribution of the means of the DV's and their linear combinations are normally distributed).
- Absence of outliers in the data.
- Homogeneity of the Variance/Covariance Matrix within each cell (allows us to pool them to create a single estimate of error variance).
- Absence of Multicollinearity (very high correlations) and Singularity (perfect correlation) in the dependent variables.

- Dependent variables are continuous in their underlying structure, and interval level.

Here is an example of a procedure that might be addressed using MANOVA:

A college professor thinks that changing text books will improve student grades in both math AND physics. She is also curious if gender is involved. She decides to randomly assign half her male and female students the old text book, and the other half the new text book.

Notice that text book version (old or new) is one independent variable, gender is a second IV, and that there are TWO dependent variables (grade in math and grade in physics). Most likely math and physics grades are correlated with one another, so improvement in one should be related to improvement in the other.

MANOVA will create a new, combined Dependent Variable, a linear combination of scores on the math and physics exam, and perform a 2 way analysis of variance on that variable. This new variable is something different from either the math scores and the physics scores, and represents their combination so as to maximize group differences. The newly created dependent variable might be called "Science Ability."

In this example, the researcher will test the main effects for the text book, the main effects for gender, and a book by gender interaction on the combined dependent variables. One main effect null hypothesis is: There will be no differences between men and women in their Scientific Ability averages, controlling for book version.

The second main effect null hypothesis is: There will be no differences between people who use the old version versus the new version of the text book in their Scientific Ability averages, controlling for gender.

The interaction null hypothesis is: There will be no differences in average scores on Scientific Ability between different genders and different versions of the text book.

As in ANOVA, post hoc tests, power, and effect sizes can all be calculated and evaluated.

The intercorrelations of physics and mathematical ability is taken into account in the creation of the new variable, rather than ignoring them and producing two separate univariate analyses. Notice that the results can have practical value in deciding which text book to order, and even (perhaps) different books for different genders!

YOUR TURN: EXERCISE 1

Think of some major variables that you might wish to study for a class assignment. Some of these may be categorical (independent variables) and some may be quantitative (dependent variables). Recall that MANOVA investigates whether or not group differences exist between multiple quantitative outcome measures considered jointly, taking into account correlations between them. Now, speculate on a potential study in your field of interest where MANOVA might be appropriate.

Discriminant Analysis

Now that we have considered MANOVA, let us look at Discriminant Analysis (DA), sometimes called Discriminant Function Analysis (DFA). In this form of multivariate analysis, our goal is prediction. However, instead of prediction of a score, as we were in multiple regression, we are interested instead in prediction of a categorical outcome. This categorical variable can be thought of as group membership.

Examples are very common: a person may survive an operation, or they may not. Someone might graduate from college, or not. A high school senior may choose to enlist in the Army, the Navy, the Air Force or the Marines. An infant may pick a blue, yellow or red toy. In all cases, a prediction of a category is the goal, not a score on an assessment. Using certain predictive variables, which themselves might be categorical (e.g., gender) or quantitative (IQ scores), can we make predictions more accurately than by chance into two (or more) groups?

So, the primary purpose of a discriminant function analysis is to use a set of variables to predict membership into two (or more) mutually exclusive and exhaustive groups.

A major goal of this analysis is to investigate the pattern of scores on the predictor variables (IV's) that predict an individual will belong to one group or another. The procedure produces a discriminant equation, where scores of the individuals in the sample are inserted, and predictions made as to which group they belong to are made. The method does this by creating an optimal linear weighted combination of the variables that allow for the highest level of accurate prediction into groups.

The initial sample of participants will have known group membership; a classification table is produced that gives a gauge as to how many "hits" (correct predictions) and "misses" (incorrect predictions) the model produces. Then, based on the discriminant equation, new participants will be classified into groups, based on their scores on the original set of predictor variables.

To summarize, group membership for a large sample is already known. Variables that are assumed to be correlated with group membership are administered to the participants in the groups. A discriminant function equation is derived that maximally separates the two groups based on the scores on the predictor variables. This equation can be applied to new individuals to predict where they are most likely to fall.

The similarity of the new and old groups is crucial for accurate prediction.

Consider the following example of Discriminant Analysis. Say that a researcher is interested in predicting if individuals entering into an online learning program will succeed (graduate from the program) or not. She hypothesizes that the following eight variables are useful in this prediction: IQ score, Undergraduate GPA, Level of familiarity with computers, Score on a validated test of Reading Ability, Gender, Marital Status, Age, and Level of Motivation.

Her dependent variable is a dichotomy: Success (graduate) or Failure (does not graduate) from the online program. She obtains scores on all the above variables for 350 people who enrolled in an online program in 2001. She divides the people into two groups: those who graduated, and those who did not.

- There are 8 predictor variables, which will result in a discriminant function equation that looks something like this: D1= aX1 + bX2 + cX3…+ hX8. In this equation, D1 is the discriminant function equation which will be used to predict a participant's group membership. Since the categorical grouping variable has only two levels, there will be only one discriminant function. If more than two levels of the grouping variable are involved, there may be more than one discriminant function. The a, b, c, etc. are the discriminant coefficients that are generated by the procedure. These coefficients allow for maximum separation between groups (SSbetween is maximized, and SSwithin is minimized).

The X1, X2, etc. are the scores of the individual on that predictor variable. So, the equation D1= aX1 + bX2 + cX3…+ hX8 can be used to classify NEW participants into groups of success (likely to graduate) or Failure (not likely to graduate). Note that if all raw scores are transformed into Z scores, a discriminant function for standard scores is calculated (just as we might do with ordinary least squares regression).

After the Discriminant Function is generated, an overall test for significance can be performed. This is a "goodness of fit" test called Wilk's Lambda, used to see if the model predicts group membership at a significantly higher rate than chance. It is similar to the concept of (1-R2 in a regression analysis), looking at the proportion of unaccounted variability in the model.

In addition to the test of the overall model, a researcher is often interested in looking at the contributions of the individual predictor variables. Which of the variables in the model seem to have the greatest predictive value, and which seem to have the least? There are several ways that this question can be answered, and these are discussed in advanced text books for graduate statistics.

YOUR TURN: EXERCISE 2

This exercise is very similar to the previous one with MANOVA. Think of some major variables that you might wish to study for a class project. Some of these may be categorical and some may be quantitative. How might a discriminant analysis be helpful in answering the research question?

Factor Analysis

The final method to be discussed in our brief introduction to multivariate statistics is known as factor analysis. Factor analysis has been around comparatively longer than MANOVA or DFA. The first really important application of Factor Analysis was conducted by Cattell, who took roughly 4500 trait labels (funny, generous, reserved, pompous) and reduced them to 200 questions that measured 16 basic components of personality (The 16 Personality Factor Questionnaire, or 16PF).

Factor Analysis is one of many multivariate methods where the primary goal is the reduction of a large amount of data into a smaller, but still meaningful number of useful constructs, or "factors." Similar to Factor Analysis is Cluster Analysis, where the goal is to classify similar object, such as cars, radio stations, consumer preferences, and so on. Factor analysis on the other hand looks at variables, specifically variables within a correlation matrix, and searches for patterns that suggest a strong similarity of some variables to others. It calls this similarity a "factor." More than one factor is usually identified in the data set, which may contain dozens or even hundreds of individual variables. The overall goal of Factor Analysis is to recreate the patterns of correlations in the matrix, but using a much smaller number of factors (rather than variables) to do so.

Factor Analysis looks at patterns of correlations among a large number of variables. Variables that are strongly correlated with other variables are assumed to measure the same underlying construct, or factor. Factors are sometimes referred to as latent variables, since they are not directly observed. The ideal structure would be to find variables that correlate highly with each other, and correlate very low with the remaining variables in the matrix. This concept is usually called simple structure.

Much like the Discriminant Function Equation we studied, the results of a Factor Analysis involve the formation of a multiple regression equation that will allow us to "understand" an individual as a function of their scores on the factors. This might be especially useful in personality research. Unlike Discriminant Analysis, Factor Analysis starts with (often) hundreds of variables, finds all of the inter-correlations among them, and proceeds to combine highly correlated variables into factors. The number of retained factors is much much less than the number of original variables, and while there is some loss of information, the advantage of simplicity outweighs this consideration.

So, factor analysis is a data reduction method designed to look for common dimensions, or latent variables, in a correlation matrix. The mathematics of factor analysis are complex, and like other multivariate procedures, requires a fair amount of reading graduate level books for a thorough understanding of the procedure. But the basic idea is simple: reduce a large amount of correlated variable to a smaller number of factors, without significant loss of information.

Factors, once extracted, can be seen as axes in a plane (two factors), in a cube (three factors), or in a multidimensional space (p factors). Each factor is orthogonal (uncorrelated) with other factors, which translates into a 90 degree angle between factors. Individual variables can be plotted in this factor space, and not surprisingly, variables that correlate (load) highly on a factor will be physically near that factor, and removed from other factors in the same factor space.

To reiterate, Factor Analysis begins by mathematically "examining" the correlation matrix generated by all of the variables. It then selects the combination of variables that explains the largest amount of variability in the matrix (This will be called "Factor 1" until a better name is created by the researcher). The second factor will consist of the combination of variables that collectively explain the greatest amount of left over variability (Factor 2). Note that Factors 1 and 2 will be uncorrelated with one another.

Additional factors will be extracted until only a negligible amount of variability is accounted for by doing so; at that point, the extraction stops. A graphical representation of the number of factors extracted, called a scree plot, can be used as a visual examination of when to stop producing additional, negligible factors. The number of factors will be significantly less than the number of variables, but will explain almost the same amount of variability in the correlation matrix as the original variables. Of course, the more factors extracted, the more closely the original correlation matrix can be reproduced. But additional factors may contribute only minute amounts of additional useful information.

Factor solutions will have to be Rotated to achieve Simple Structure. This means that the variance is redistributed among the factors, but with the intention of having as many variables "load" on that factor. Rotation is both a mathematical and geometric process that makes the final solution more interpretable. Multiple methods of rotation can be chosen. Simple Structure means maximizing high correlations and minimizing small ones on separate factors.

Here is an example of a research design that might use factor analysis: A researcher has access to a questionnaire with 100 separate questions presumed to measure personality. The researcher feels that many of these questions are redundant, and that a much smaller number of factors probably accounts for most of the information in the 100 questions.

Factor Analysis can reduce the number of questions (variables) into a much smaller number of factors; these factors reflect commonality (correlation) between several of the original questions (variables) in the set. Think of factors as basic "dimensions" of a construct; in some ways it is similar to breaking down a molecule into its component atoms. Factors, like atoms, are more basic, fundamental units.

Participants can be assessed on the newly created set of factors, rather than the original set of 100 questions. The number of factors will be considerably less than the number of variables. The 100 questions might, for example, be reduced to 8 factors; then each person is scored on each of the 8 factors. The results of this are an understanding of the individual on these factors. Highly correlated questions are "lumped together" into a factor. The first factor explains the most variability of the correlation matrix; the second the second most remaining variance, and so on.The unique pattern of scores on the factors "define" the individual. Factors are given descriptive names, depending upon the variables that make them up in the first place.

YOUR TURN: EXERCISE 3

Think of an example in your area of research interest where a factor analysis might be an appropriate method of investigation.

In summary, Multivariate Statistics are extremely powerful methods to investigate important research questions and hypotheses. These methods are very complex in their details, but quite understandable in their basic intent. However, you do not need to understand the underlying mathematics to use and appreciate them!

Whatever form of statistical analysis one uses, it is imperative that careful thought be given to the research question. Care in the design of the study is much more important than any fancy statistics, and of course professional looking tables and graphs will not save a poorly designed study. Statistics are useful and necessary tools, but only as good as the logic, flow, variables and materials that the design offers.

3

QA 276.12 .T78 2016

Trunk, Barry,

Ok, i've signed up for
statistics. now what?